Fuel Cells Symposium 2016

Held at the 33rd Annual International
Battery Seminar & Exhibit 2016

Fort Lauderdale, Florida, USA
24 March 2016

ISBN: 978-1-5108-2255-9

Printed from e-media with permission by:

Curran Associates, Inc.
57 Morehouse Lane
Red Hook, NY 12571

Some format issues inherent in the e-media version may also appear in this print version.

Printed by Curran Associates, Inc. (2016)

For permission requests, please contact International Battery Seminar
at the address below.

International Battery Seminar
250 First Avenue, Suite 300
Needham, MA 02494
USA

Phone: (561) 367-0193
Fax: (561) 367-8429

info@powersources.net

Additional copies of this publication are available from:

Curran Associates, Inc.
57 Morehouse Lane
Red Hook, NY 12571 USA
Phone: 845-758-0400
Fax: 845-758-2633
Email: curran@proceedings.com
Web: www.proceedings.com

CO-LOCATED EVENT SYMPOSIUM 2: MARCH 24, 2016

Knowledge Foundation's 16th International

FUEL CELLS

Next Generation Fuel Cells for Commercial, Military & Industrial Applications

7:30 am **Symposium Registration and Morning Coffee**

8:00 **Organizer's Opening Remarks**
Craig Wohlers, Executive Director, Conferences, Cambridge EnerTech

8:05 **Chairperson's Opening Remarks**
Allison Fisher, Ph.D., Visiting Professor, University of Central Florida

8:10 **Featured Presentation: Fuel Cell and Hydrogen Energy - The Future is Now** **1**
Morry Markowitz, President, Fuel Cell & Hydrogen Energy Association

FUEL CELLS MATERIALS, DESIGN & MODELING

8:40 **Novel Anion Exchange Membranes for Hydrogen and Hydrazine Alkaline Fuel Cells** **N/A**
Kenji Miyatake, Ph.D., Professor, Clean Energy Research Center, University of Yamanashi, Japan

9:10 **Advances in MEA Development: Automotive Application, Low PGM, Durability, Materials by Design Approach** 10
Madeleine Odgaard, CEO, IRD Fuel Cells, LLC, Denmark

9:40 **Coffee Break with Poster Viewing**

ADVANCED FUELS AND INNOVATIVE APPROACHES

10:40 **An Appreciation of Solid Hydrogen Storage and Release** **24**
Allison Fisher, Ph.D., Visiting Professor, University of Central Florida

11:10 **Enjoy Lunch on Your Own**

ADVANCED FUEL CELLS FOR TRANSPORTATION APPLICATIONS

1:20 pm **Chairperson's Remarks**
Andrew Burke, Ph.D., Institute of Transportation Studies, University of California, Davis

1:25 **Development of Heavy Duty Fuel Cell Modules for Mass Transport Applications** **N/A**
Mauricio Blanco, Ph.D., Research Engineer, Strategic Technology Planner, Ballard Power Systems

1:55 **Natural Gas as a Bridge to Hydrogen Fuel Cell Light-Duty Vehicles** **32**
Andrew Burke, Ph.D., Institute of Transportation Studies, University of California, Davis

2:25 **Fuel Cells for Unmanned Systems Propulsion** **41**
James C. Sisco, Ph.D., Principal Engineer, Protonex Technology Corporation

2:55 **Refreshment Break with Poster Viewing**

3:30 **Methanol Vapor Fuel Cell for Optimum Light Weight Power** **47**
Walt Robb, Ph.D., CEO, MeOH Power

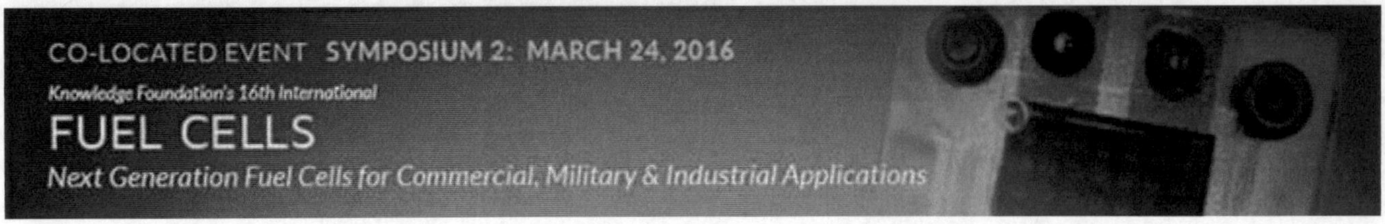

4:00 **Fuel Cell and Battery Hybrid for Energy Storage - Battery for Rapid Response and Fuel Cells for Range Extension for Storage** 55
Pinakin Patel, Director, Research & Special Systems, FuelCell Energy, Inc.

4:30 **Fuel Cells for Industrial Vehicles 68**
Gus Block, Director, Nuvera Fuel Cells

5:00 **Close of Symposium & Close of International Battery Seminar & Exhibit**

SPEAKER LISTING

Dr. Mauricio Blanco
Research Engineer
Ballard Power Systems, Inc.
9000 Glenlyon Pkwy.
Burnaby, BC V5J 5J8
Canada

Gus Block
Director
Marketing & Government Affairs
Nuvera Fuel Cells
129 Concord Rd.
Bldg. 1
Billerica, MA 01821

Chris D'Couto
President & CEO
Neah Power Systems, Inc.
22118 20th Ave. SE
Ste. 142
Bothell, WA 98021

Allison M. Fisher
Senior Research Scientist
Space Life Sciences Lab
Cella Energy US, Inc.
505 Odyssey Way
Kennedy Space Center
Exploration Park, FL 32953

Morry Markowitz
President & Executive Director
Fuel Cell & Hydrogen Energy Association
1211 Connecticut Ave. NW
Ste. 600
Washington, DC 20036

Kenji Miyatake
Professor
Clean Energy Research Center
University of Yamanashi
Clean Energy Research Center
4 4 37 Takeda Kofu
Yamanashi, 400-8510
Japan

Madeleine Odgaard
CEO
IRD Fuel Cells LLC
8500 Washington St. NE
Ste. B1
Albuquerque, NM 87113

Dr. Pinakin Patel
Director
Special Systems & Research
FuelCell Energy, Inc.
3 Great Pasture Rd.
Danbury, CT 06810-8153

Dr. Walt L. Robb
CEO
MeOH Power
874 Old Albany Shaker Rd.
Latham, NY 12110

Dr. James C. Sisco
Principal Engineer
Protonex Technology Corp. a Ballard Co.
153 Northboro Rd.
Southborough, MA 01772

Fuel Cell &
Hydrogen Energy
Association

Fuel Cells and Hydrogen Energy: The Future is Now

Morry Markowitz
International Battery Seminar
Ft. Lauderdale, Florida
March 24, 2016

About FCHEA

- The Fuel Cell and Hydrogen Energy Association (FCHEA) is the trade association for the fuel cell and hydrogen industry located in the United States

Our Mission

- Our mission is to advance the commercialization of, and promote the markets for, fuel cell and hydrogen energy technologies

Fuel Cell &
Hydrogen Energy
Association

Our Members

FCHEA's Priorities

- Build and Strengthen the Fuel Cell/ Hydrogen Brand
- Enhance Government Affairs Activities
- Accelerate Hydrogen Infrastructure Development / Fuel Cell Deployment
- Create Environment for Increased Product Sales

Fuel Cell & Hydrogen Energy Association

FCHEA Working Groups and Committees

- Government Affairs
- Communications and Marketing
- Transportation
- Stationary Power
- Portable Power
- Solid Oxide Fuel Cell

Fuel Cell & Hydrogen Energy Association

Today's Fuel Cell Markets

- **Transportation** – cars, buses, scooters, material handling
- **Hydrogen Fueling** – distribution, generation, station components
- **Stationary Power** – primary power, distributed generation, back up power
- **Portable Power** – battery chargers, remote/off-grid, recreation, military

Why Fuel Cells?

- Low/zero emissions
- Resilient/reliable
- Scalable to any power need
- Fuel flexible – operation on conventional or renewable fuels
- Can partner with batteries, solar, wind, and other renewable technologies

Fuel Cell Electric Vehicle Developments

- Hyundai Tucson Fuel Cell
  - Leasing in Southern California
- Toyota Mirai FCEV
  - Eight dealerships selling and leasing in California
- Honda Clarity
  - U.S. sales to start mid-2016
- Automaker Partnerships
  - GM/Honda
  - Daimler/Ford/Nissan-Renault
  - Toyota/BMW

California H2 Station Map

- ▼**Retail** – 11
- ▼ **In Development**
  - Q1 16– 8
  - Q2 16– 12
  - Q3 16 – 1
  - Q4 16 – 9
- **Other Open**
  - ▼Bus – 3
  - ▼Non-retail – 5

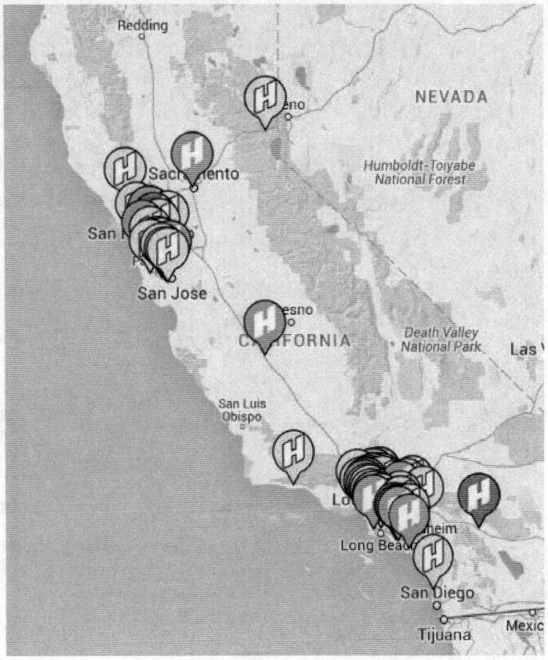

Northeast H2 Station Rollout

New York, New Jersey, Connecticut, Rhode Island, and Massachusetts

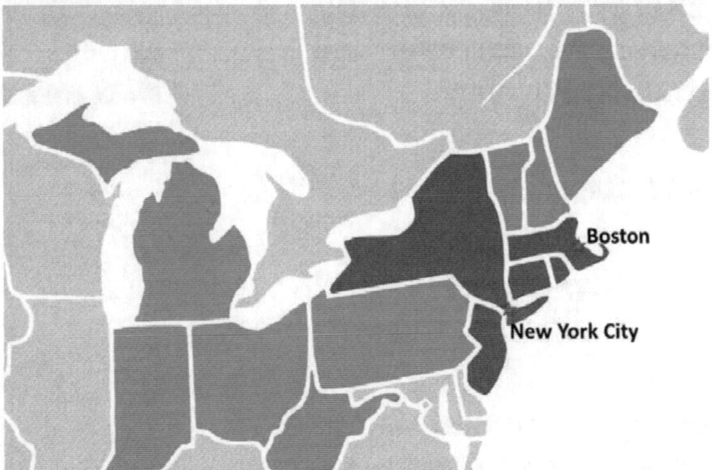

U.S. Material Handling Fuel Cell Industry

Leading Fuel Cell Forklift Companies

Plug Power   Nuvera Fuel Cells

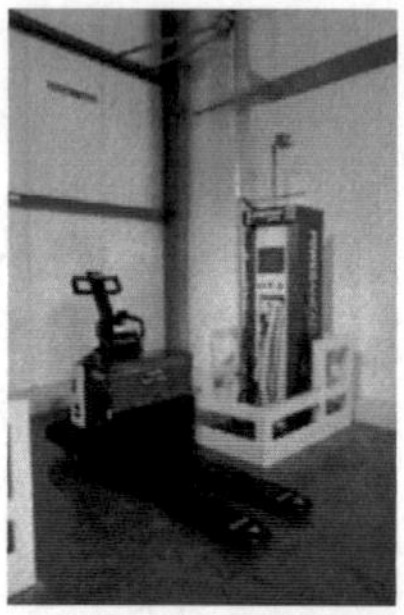

7,700+ fuel cell-powered forklifts are deployed in the U.S. today

Fuel Cell Material Handling Customers

Kroger Walmart The Home Depot Lowe's

Good things come from Sysco Wegmans P&G FedEx

Honda The Power of Dreams The Coca-Cola Company BMW

Fuel Cell & Hydrogen Energy Association

U.S. Stationary Fuel Cell Industry

- Home to leading fuel cell companies – Bloom Energy, FuelCell Energy, LG Fuel Cell Systems, Doosan Fuel Cell America, Plug Power

- >230+ MW currently installed in U.S.
- > 6,000 backup power fuel cells operating at telecom, railroad sites

- New customers/markets include utilities, micro grids, energy storage

Fuel Cell & Hydrogen Energy Association

Sample of U.S. Stationary Fuel Cell Customers

FedEx Panasonic verizon Google Apple

ebay inc COX ENTERPRISES HONDA The Power of Dreams delmarva power

Adobe The Coca-Cola Company NOKIA Walmart

Sprint AT&T CSX metroPCS

Fuel Cell & Hydrogen Energy Association

H_2USA

H_2USA

H_2USA Mission Statement

The mission of H_2USA is to promote the commercial introduction and widespread adoption of FCEVs across America through creation of a public-private collaboration to overcome the hurdle of establishing hydrogen infrastructure.

H_2USA

Current Signatories on the Letter of Understanding

H_2USA

H_2USA Organization Chart

H2USA

Thank you!

Fuel Cell and Hydrogen Energy Association
1211 Connecticut Avenue, Suite 650
Washington, DC 20036
202-261-1331
www.fchea.org

Fuel Cell & Hydrogen Energy Association

Fuel Cell Symposium 2 @ International Battery Seminar & Exhibit

March 24th 2016, Fort Lauderdale Convention Center, Fort Lauderdale, FL

Advanced MEA Development: Automotive Applications, Low PGM, Durability, Materials by Design Approach

Madeleine Odgaard, IRD Fuel Cells

IRD
POWERING PEOPLE ANYWHERE

2

IRD Fuel Cells – The Core Inside

Our vision is to provide energy to people through deployment of fuel cells

Independent strategic provider of cutting edge "core" for fuel cells and electrolyzers

Striving for research excellence in our core technologies

Quest for customer centricity – reduce cost of energy

Our Promoters TREFOR

IRD
POWERING PEOPLE ANYWHERE

March 24th 2016, Fort Lauderdale Convention Center, Fort Lauderdale, FL

3

IRD Fuel Cells – The Core Inside

Manufacturing Platform: Denmark & USA

Technology:

- Customized prototype manufacturing capability
- Unique high quality CCM / MEA manufacturing technology with
- Press molding production facility for flow plates for high volume
- MEA production facilities both US and Denmark

Collaboration with international research groups

Strategic long term partnerships:

- R&D contracts / co-development with customers
- Customized prototype manufacturing,
- Long-term supply agreements at volume manufacture

IRD
POWERING PEOPLE ANYWHERE

March 24th 2016, Fort Lauderdale Convention Center, Fort Lauderdale, FL

4

Industrial Application of PEM FC

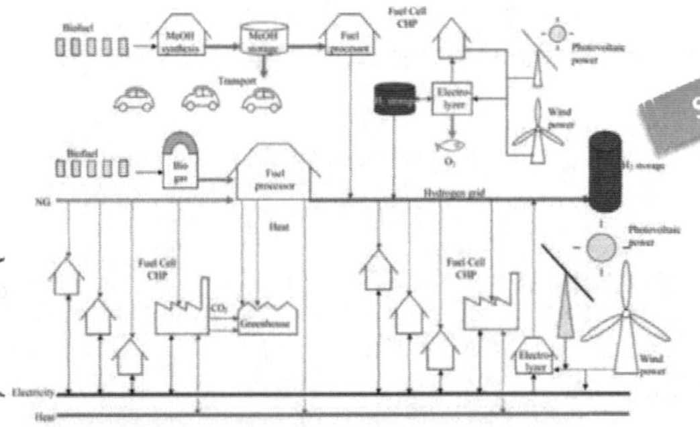

Concept of the hydrogen society

Special Service vehicles

www.nuvera.com

www.proton-motor.com

FC elctric busses

US Hybrid AC Transit bus

Transport for London
www.ballard.com

FC electric vehicles

IRD
POWERING PEOPLE ANYWHERE

March 24th 2016, Fort Lauderdale Convention Center, Fort Lauderdale, FL

5

R&D Activities – Targets Automotive MEAs

Table 3.1.1.1 State-of-the-art and future targets for Transportation

Application	Parameter	Unit	2012	FCH-JU target 2017	2020	2023
Fuel cell electric passenger cars	Specific FC system cost Assumed number of units (per year) as cost calculation base	€/kW	>500	150 20 000	100 50 000	75 100 000
	FC vehicle cost (C-segment)	k€	200	70	50	30
	Tank-to-wheel efficiency (vehicle in New European Drive Cycle)	%	40	42	45	48
	Availability	%	95	98	98	99
	FC system Lifetime	hours	2500	5000	6000	7000
Fuel cell electric buses	Specific FC system cost	€/kW	< 3500	< 1800 750	1000 500	800 400
	FC Bus System Lifetime	hours	10000	15000 2 x 8000	20000 2 x 10000	25000 2 x 12500
	FC Bus cost	k€	1300	700	650	500
	Fuel consumption (vehicle, average of SORT1 and SORT2 cycle)	kgH2/100km	9	8.51	8	7.59
	Availability	%	85	90	95	99
	Assumed number of units (per year) as cost calculation basis			< 50	200	> 500
Hydrogen Storage	Hydrogen storage system cost	€/kg H2	>1000	800	600	500
	Volumetric capacity (H2 tank system)	kg/l	0.02	0.022	0.023	0.025
	Gravimetric capacity (H2 tank system)	%	< 4	4	5	6
Material handling	Specific FC system cost	€/kW @10 kW	4 000	< 1500	< 1200	< 1000

Ref: FCH-JU Multi Annual Workplan 2014-2020

Table 3.4.5 Technical Targets: Membrane Electrode Assemblies

Characteristic	Units	2015 Status	2020 Targets
Cost[a]	$ / kW_{net}	17[b]	14
Durability with cycling	Hours	2,500[c]	5,000[d]
Startup/shutdown durability[e]	Cycles	–	5,000
Performance @ 0.8 V[f]	mA / cm^2	240[g]	300
Performance @ rated power[h] (150 kPa abs)	[illegible]	[illegible]	[illegible]
Robustness (cold operation)[i]	[illegible]	1.09	0.7
Robustness (hot operation)[j]	[illegible]	[illegible]	0.7
Robustness (cold transient)[m]	[illegible]	[illegible]	0.7

Performance, Quality & Capability

Ref. DoE Roadmap FOA-0001412

Cost: materials, process

IRD
POWERING PEOPLE ANYWHERE

March 24th 2016, Fort Lauderdale Convention Center, Fort Lauderdale, FL

6

EU Project IMMEDIATE -

IMMEDIATE

FCH JU 303466

Objective, Targets

The prime focus of the work is to develop high performing MEAs aimed for automotive applications ***through material R&D & process optimization***.

Develop a medium temperature PEM Membrane Electrode Assembly that will fulfil the OEM requirements with respect to ***cost, performance and durability*** and at the same time is a significant step towards the ultimate goal to reduce ***Pt-loadings <0.15 g/kW*** at >55% efficiency

Characteristic	Targets
PGM Total content	< 0.15 g/kW
Lifetime (h)	> 5000 hr
Membrane proton conductivity	> 100 mS/cm @ 25 %RH & 120 °C
BoL Power Density	> 1 W/cm2 @ U_{Cell}=0.68V, nominal operating conditions
EoL Power Density	> 0.9 W/cm2 @ U_{Cell}=0.68V, nominal operating conditions

IRD POWERING PEOPLE ANYWHERE

March 24th 2016, Fort Lauderdale Convention Center, Fort Lauderdale, FL

7

MEA - Materials by Design /

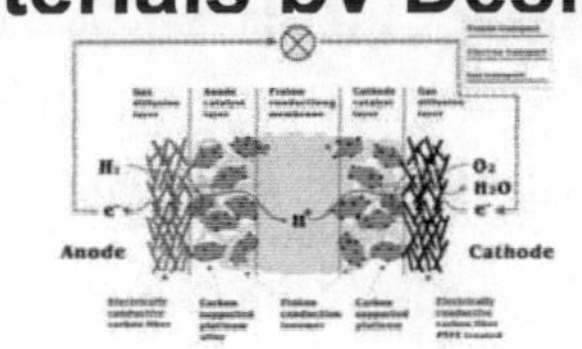

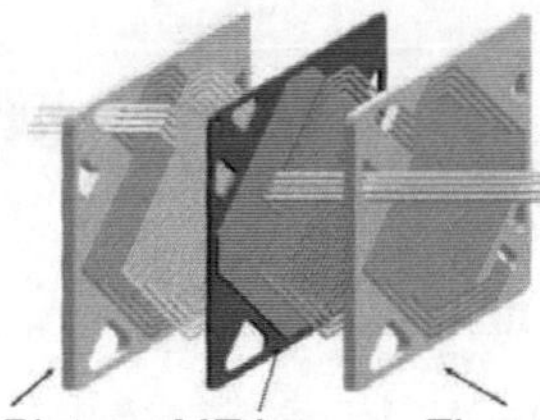

Approach

Material R&D & process optimization to improve and incorporate in new innovative materials:

- Novel catalyst support materials with tailored surfaces and pore structure will be developed, tuned to optimise both the Pt utilisation and the mass transport of the reactants and products
- Innovative low equivalent weight cross-linking PFSA ionomers and catalysts that survive the high temperature and low RH by improvement of the oxidation durability and the proton conductivity
- GDL materials and MPL will be further optimised for improved water and gas management
- The mutual compatibility of materials and their durability shall be verified by assembling high performance MEAs for benchmarking purposes and for testing

IRD POWERING PEOPLE ANYWHERE

March 24th 2016, Fort Lauderdale Convention Center, Fort Lauderdale, FL

8

MEA - Materials by Design Approach

- Aim: **Pt/C catalysts with high durability**, high performance, high Pt utilization in electrode structures.
- Support materials: Six experimental carbons with intermediate surface area (ca. 400 $m^2 g^{-1}$); great resistance towards electrochemical corrosion; obtained from different production processes and post treatments; varying morphology and surface chemistry.
- **Correlations between support characteristics and stability/catalyst activity**

MEA - Materials by Design Approach

RDE (Rotating disc electrode analysis) measurements to initially screen the catalyst activity

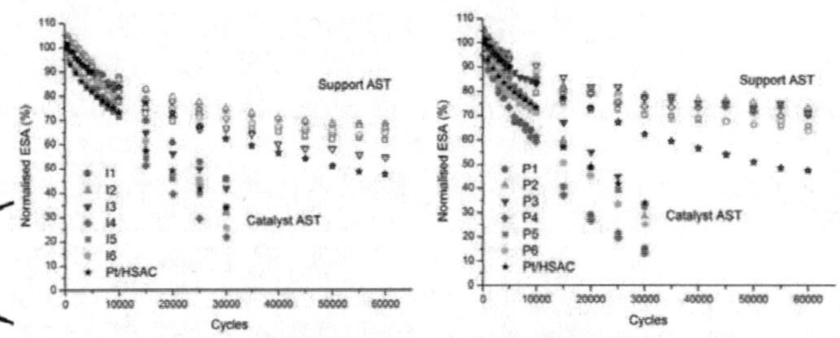

Test conditions:
0.10 M $HClO_4$
N_2 saturated
34 ± 3 °C

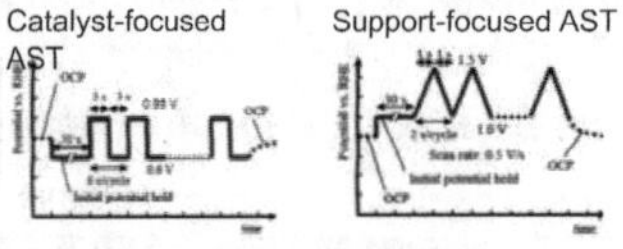

Ohma A, et al. ECS Trans 2011;41:775–84

➢ Superior support durability compared to SoA

MEA - Materials by Design Approach

Carbon	Hydrophobic character	Hydrophilic character	Hydrophilic-hydrophobic ratio	Aromatic character
	$D_{ads}H$ ($J \cdot g^{-1}$)	$D_{ads}H$ ($J \cdot g^{-1}$)	$D_{ads}H / D_{ads}H \cdot 100$	$D_{ads}H$ ($J \cdot g^{-1}$)
CB1	54.8	1.6	2.9	47.2
CB2	42.2	2.4	5.7	56.1
CB3	45.2	2.9	6.4	49.7
CB4	48.6	6.0	12.3	74.3
CB5	40.4	9.9	24.5	14.3
CB6	38.6	-	-	62.7

Carbon support	Oxygen content (%)
CB1	0.2
CB2	0.21
CB3	0.19
CB4	0.17
CB5	0.57
CB6	0.054

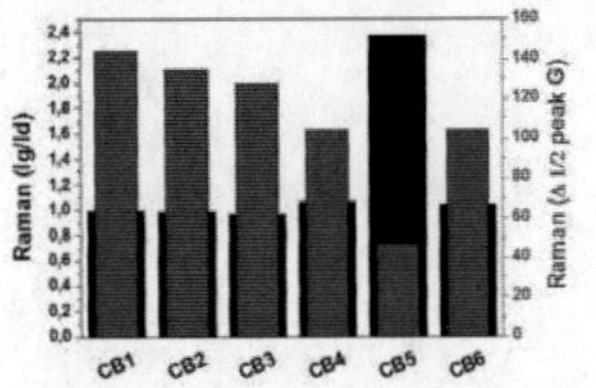

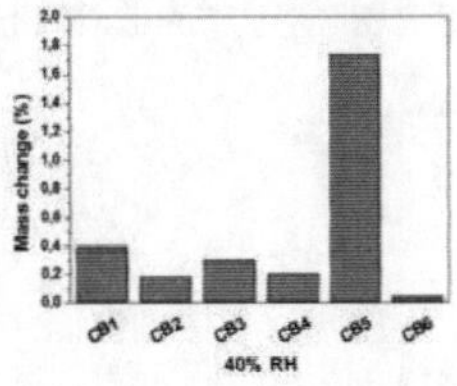

- Excellent concordance between different techniques: hydrophilic /hydrophobic ratio, oxygen content, water uptake
- Carbon with lowest hydrophilic/hydrophobic ratio => most stable catalyst on voltage cycling to 1.6 V
- next step is to evaluate how that translates to MEA Fuel Cell performance.

IRD
POWERING PEOPLE ANYWHERE

MEA - Materials by Design Approach

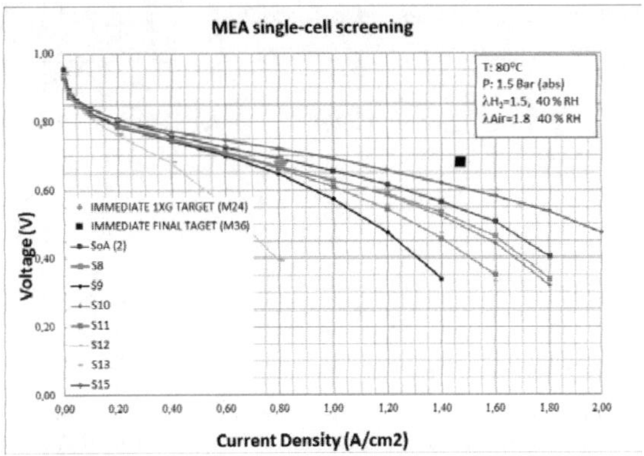

- Correlation between mass transport losses and lower hydrophobic character of carbon for partial saturated (RH) FC operation conditions

IRD
POWERING PEOPLE ANYWHERE

MEA – Electrode design and optimization

Catalyst integration includes: Ink formulations and deposition methods for spraying optimal MEA's.

- Ink & Coating development
  - composition – electrocatalyst, binders
  - stability
  - coatablity – several routes to fabricate catalyst layers

Inks -

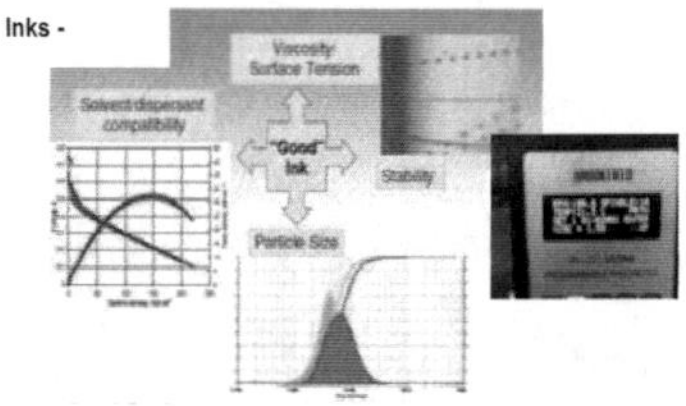

➢ Optimal electrode/membrane integrity
➢ Link detailed catalyst morphological properties to electrode optimization and stability
➢ Understand how the catalyst layers perform in conditions pertinent to different applications

March 24th 2016, Fort Lauderdale Convention Center, Fort Lauderdale, FL

MEA – Testing

All data are representative and generated under typical test conditions in 50 cm^2 fuel cell hardware and Test equipment from Fuel Cell Technologies, Inc.

Comprehensive testing of the electrode materials, MEA performance and durability of the various MEA s designs

Fuel Cell Protocols & Targets established and continuously updated for components, MEAs and cells

Different protocols for different organizations JCH-JU (working groups), US DOE, US DRIVE, FCCJ, JARI tend to have similarities.

PGM (Pt stability), carbon corrosion, membrane stability (chemical/mechanical), MEA performance and durability using Accelerated Stress Test (AST).

Challenge is *to understand* the interplay between MEA components as well as cell hardware upon *MEA* AST testing for stability evaluation and qualification

March 24th 2016, Fort Lauderdale Convention Center, Fort Lauderdale, FL

MEA – Testing

EU Harmonized Test Protocols for PEMFC MEA Testing in Single Cell Configuration for Automotive Applications

All data are representative and generated under typical test conditions in 50 cm² fuel cell hardware and Test equipment from Fuel Cell Technologies, Inc.

Materials → MEA Testing in Single Cells → Stacks → Systems

Accelerated Stress Test
Ex-situ & In-situ ⇒ Begin-of-Life
Accelerated Stress Test ⇒ Validation in relevant environment

Automotive Reference Conditions	Elevated Temperature
Power Density @ 0.68V [W/cm2]	Power Density @ 0.68V [W/cm2]
2.5/2.3 Bar (abs): 80°C: 50/30 % RH	1.5/1.5 Bar (abs):95°C: 30/30 % RH

MEA – Electrode design and optimization

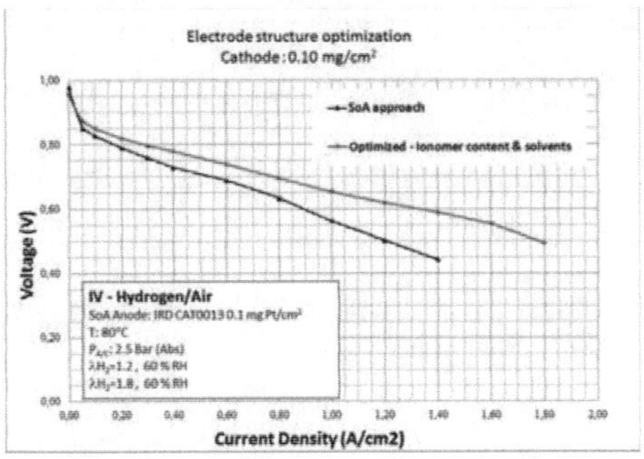

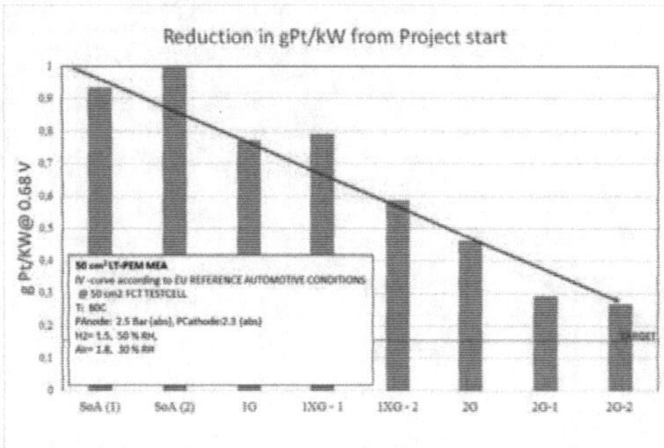

⇒ Different electrode design for the same batch catalyst
⇒ Different ionomer & solvent system for optimal performance
⇒ Improved Pt-catalyst utilization by novel materials integration & engineering of structures

MEA – Electrode design and optimization

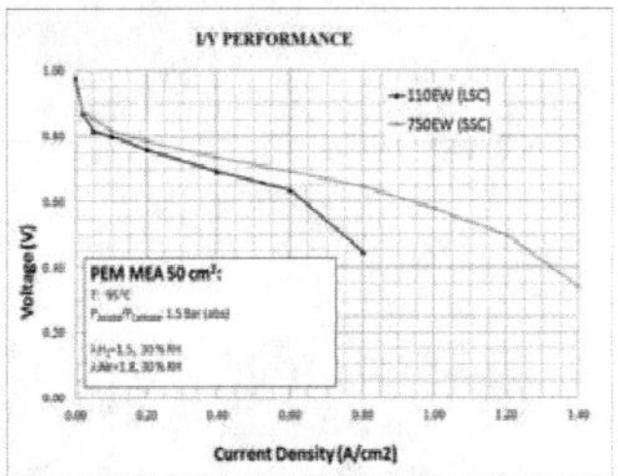

LSC vs SSC

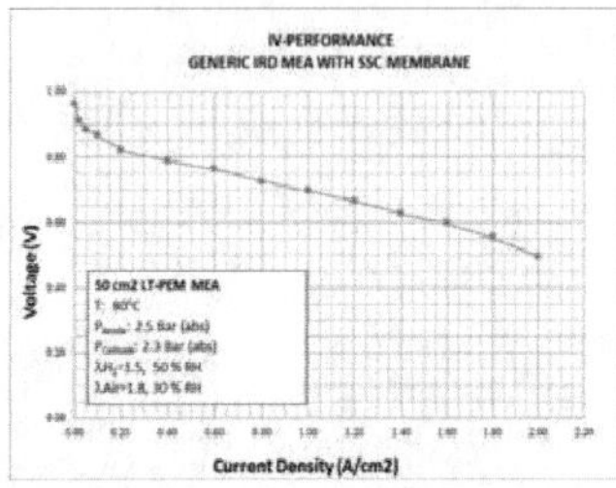

Low PGM MEA

- SSC.PFSA membranes show promising results - exhibit high proton conductivity at elevated temperatures and lower relative humidity
- Combined with enhanced GDL and low-PGM loading electrodes these MEA show promising BoL performance
- Single-cell AST test a good way to initially screen stability
  ⇒ *Link to performance and interplay in automotive FC stack/system ???*

IRD
POWERING PEOPLE ANYWHERE

March 24th 2016, Fort Lauderdale Convention Center, Fort Lauderdale, FL

17

MEA – Electrode design and optimization

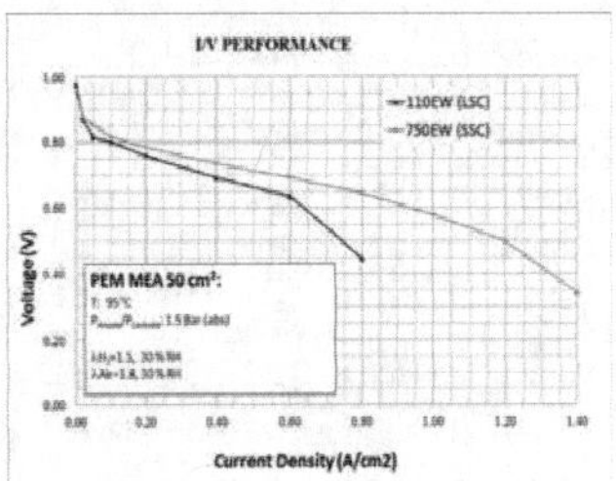

LSC vs SSC

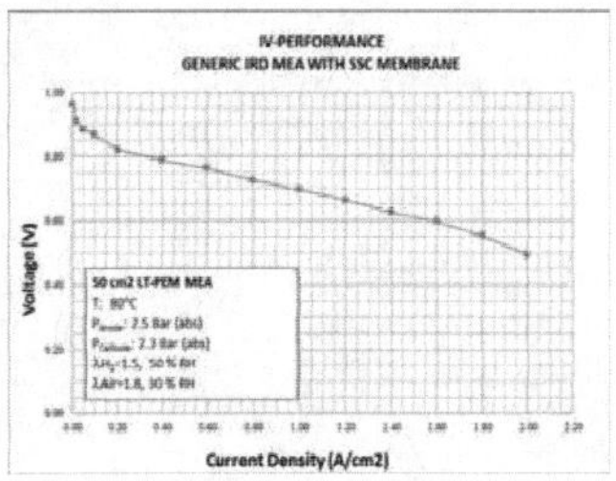

Low PGM MEA

- SSC.PFSA membranes show promising results - exhibit high proton conductivity at elevated temperatures and lower relative humidity
- Combined with enhanced GDL and low-PGM loading electrodes these MEA show promising BoL performance

IRD
POWERING PEOPLE ANYWHERE

March 24th 2016, Fort Lauderdale Convention Center, Fort Lauderdale, FL

18

MEA – Low Precious Group Metal (PGM)

Test according to:
Harmonized test protocols for the development of PEMFC single cells for FCH-JU supported projects for automotive applications

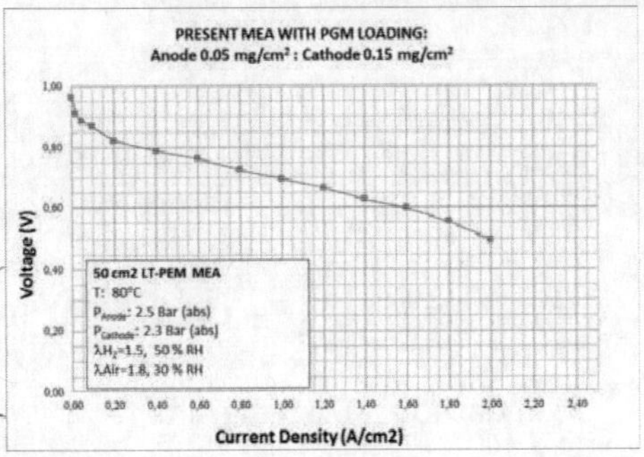

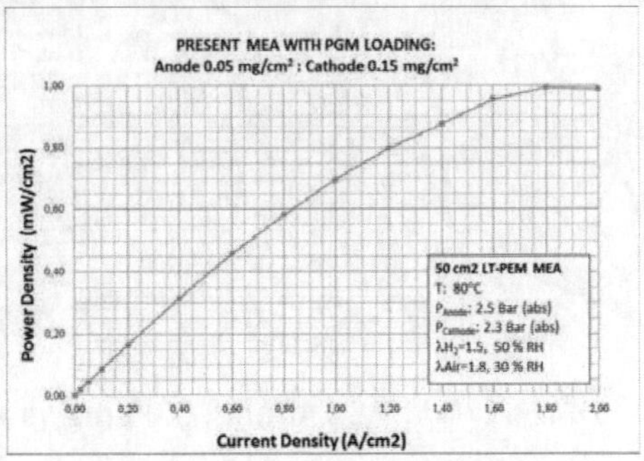

Power density 0.75 W/cm2 @ 0.68 V [auto ref.] ⇒ 0.27 gPt/kW
Max. Power density 1.0 W/cm2 ⇒ 0.20 gPt/kW

March 24th 2016, Fort Lauderdale Convention Center, Fort Lauderdale, FL

19

MEA – Low Precious Group Metal (PGM)

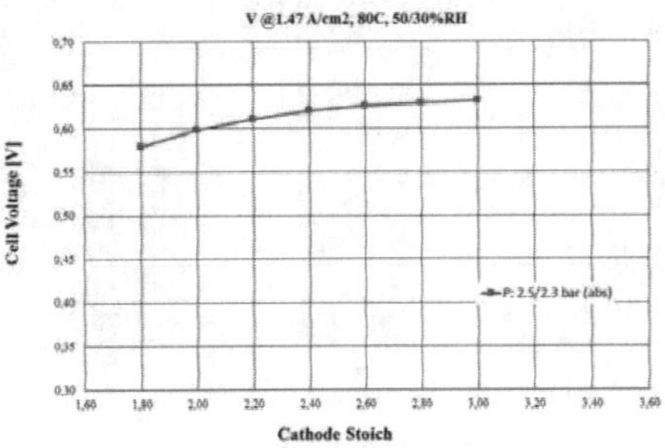

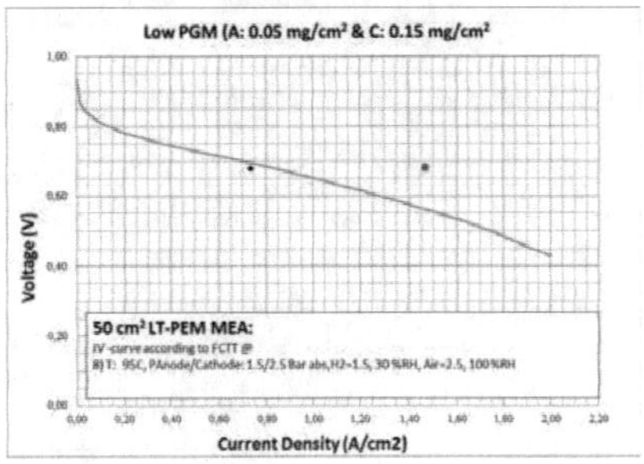

Max. Power density 0.94 W/cm2 @ 0.68 V [auto ref.]
⇒ 0.21 gPt/kW

Power density 0.448 W/cm2 @ 0.68 V
[95°C, 30%RH, 1.5 Bar $_{abs}$]
⇒ 0.45 gPt/kW

March 24th 2016, Fort Lauderdale Convention Center, Fort Lauderdale, FL

20

MEA – Durability

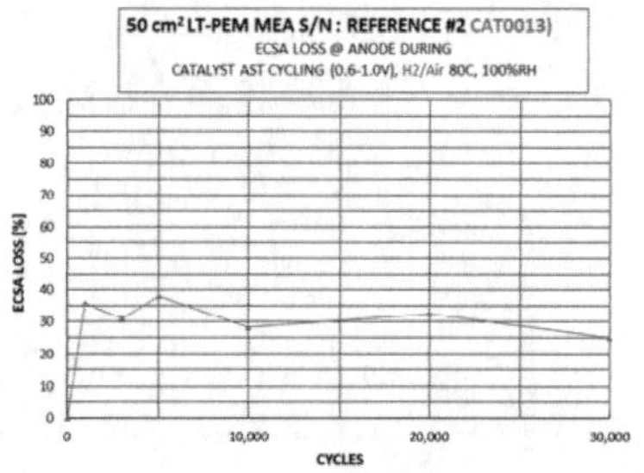

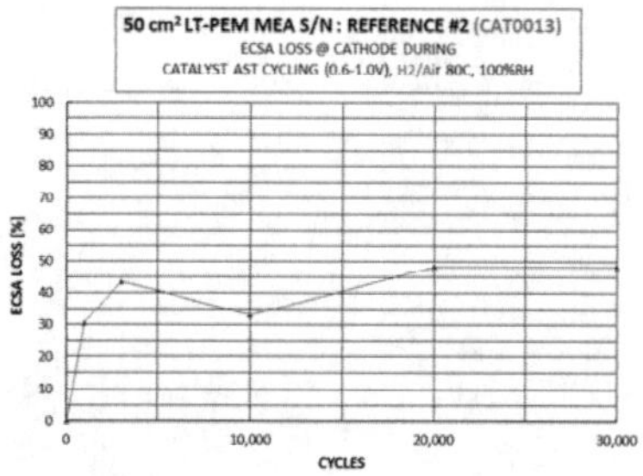

AST Anode: 0.05 mg/cm^2 Cathode: 0.15 mg/cm^2

- ➢ MEA Single-cell catalyst test using the same protocol as RDE pre-screening

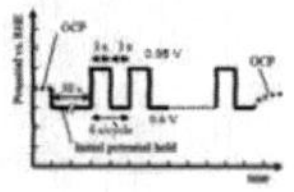

IRD
POWERING PEOPLE ANYWHERE

MEA – Durability

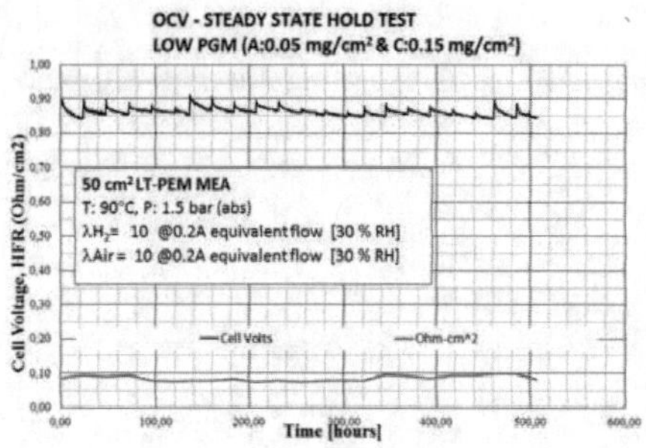

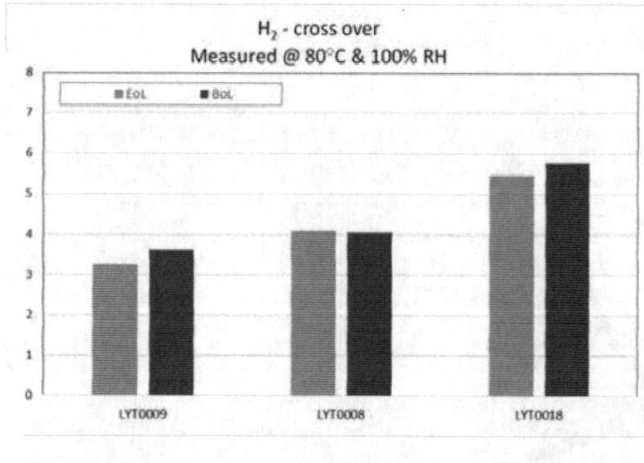

- ➢ Stable H_2 crossover during AST steady state

- ➢ Single-cell AST test a good way to initially screen stability
  - ⇒ *Link to performance and interplay in automotive FC stack/system ???*

IRD
POWERING PEOPLE ANYWHERE

MEA – Durability

Fuel Cell Dynamic Load

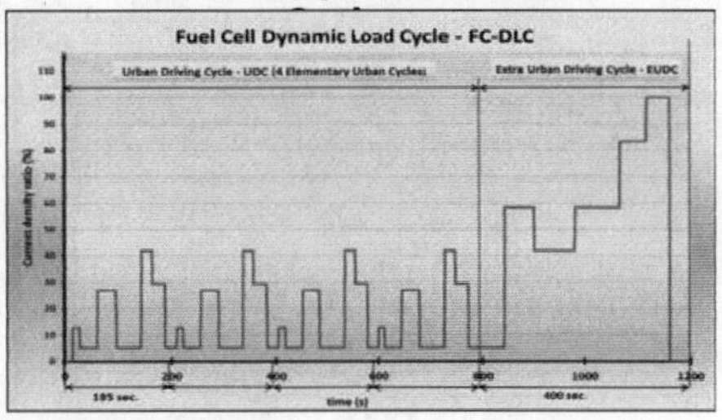

Repeat to cycle until the cell voltage, (recorded at 100% current density has decreased by 10% of its initial value.

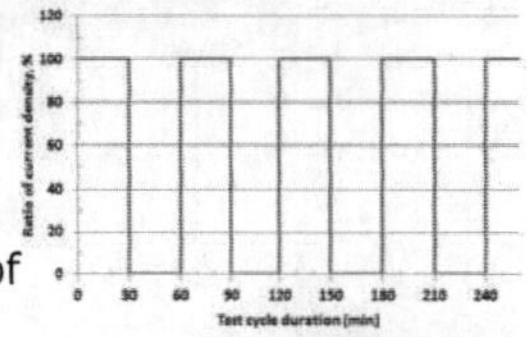

MEA – Durability

- Duty cycle tests (synthetic city bus cycle)

- These tests aim to reproduce the actual usage of the FC stack in a fuel cell system of the chosen application (city bus)

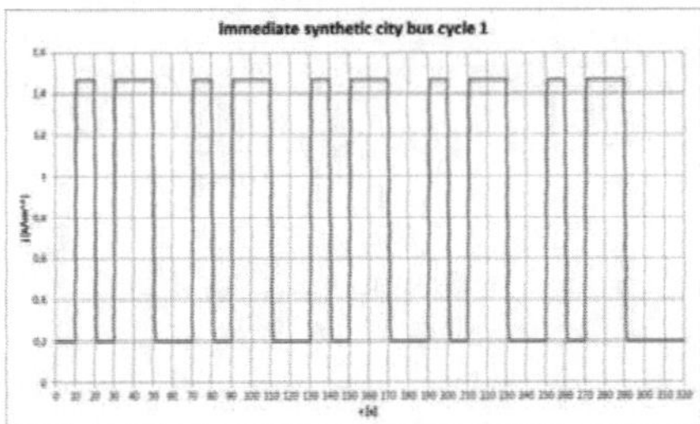

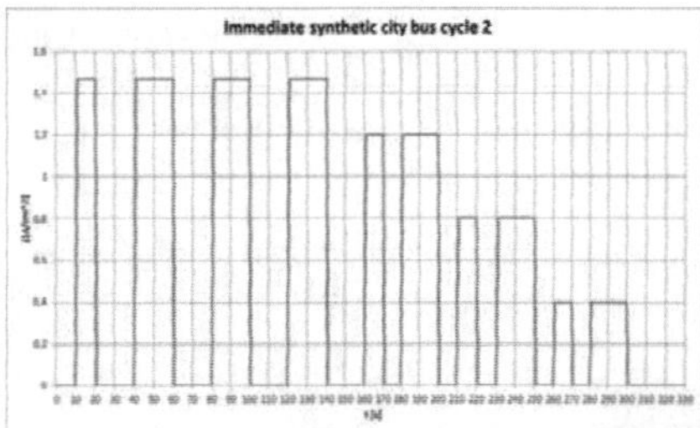

The cycles implemented, should be run repeatedly, without interruption between repetitions for as long as possible up to 15h/day until a total of 500 operating hours is reached.

MEA – Durability

Anode			Cathode			
Stoichometry	% RH	BackPressure (bar abs)	Stoichometry	% RH	BackPressure (bar abs)	Cell Temp.
1.5	40	1.5	2.0	40	1.5	80

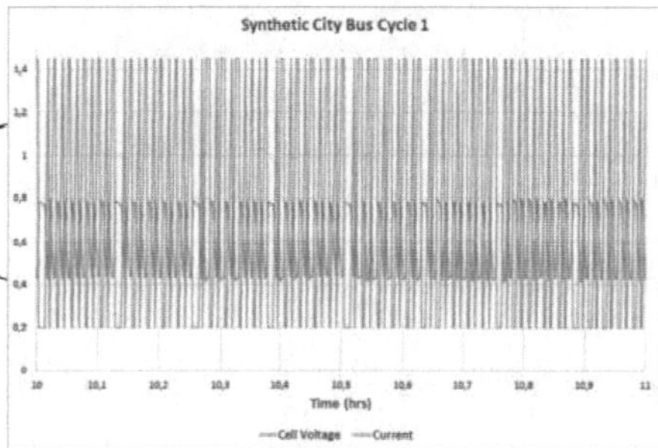

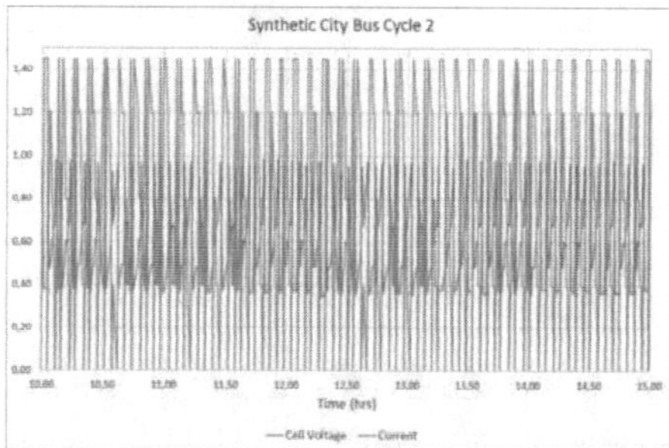

MEA – Durability

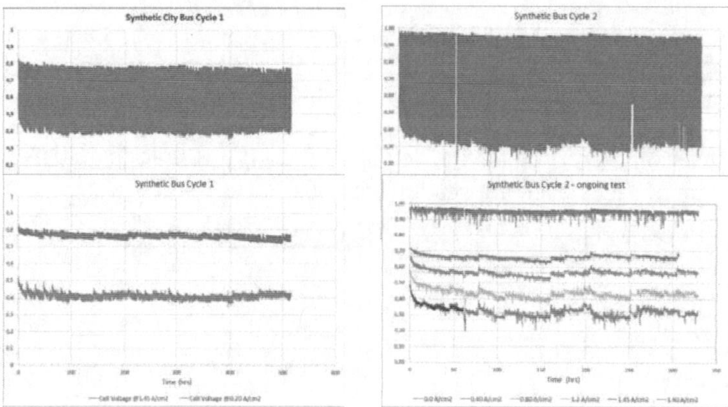

Current Density [A/cm2]	Loss [%] @ 500 hours
0.20	4.4
1.45	5.2

Current Density [A/cm2]	Loss mV [%] @ 330 hours
0.00	1.4
0.40	0.3
0.80	1.5
1.20	7.4
1.40	5.6

- Fail criteria: Loss of 10% of its initial value.

Summary

- Screening of a family of carbons for their properties as catalyst supports has been performed
- The up-scaled catalysts had considerably better stability, especially support durability, was demonstrated for the novel catalysts
- Optimization of electrode design, i.e composition and processes is unique for every combination of system operating requirements and materials set
- Advanced catalyst integration similar requires re-engineering
- This is true for ultra low loading electrodes
- Durability well understood at individual component level - Challenge is *to understand* the interplay between individual MEA components as well as cell hardware upon MEA AST testing for durability evaluation

⇒ *Link to performance and interplay in automotive FC stack/system ???*

IRD

March 24th 2016, Fort Lauderdale Convention Center, Fort Lauderdale, FL

Acknowledgements

The author acknowledge Mikkel.J. Larsen (IRD) and Debbie Schlueter (IRD) for testing and results

The research leading to these results has received partial funding from the European Union's Seventh Framework Programme for the Fuel Cells and Hydrogen Joint Undertaking under Grant Agreement No. 303466 IMMEDIATE.

Institute of Chemical Process Fundamentals of the Czech Academy of Sciences, v. v. i. (CZECH REP.)

Centre national de la recherche scientifique (FRANCE)

Shanghai Jiao Tong University (CHINA)

Volvo Technology AB (SWEDEN)

Imerys Graphite & Carbon Switzerland Ltd. (SWITZERLAND)

European Commission Directorate-General JRC, Joint Research Centre (BELGIUM)

SGL Carbon GmbH (GERMANY)

FuMA-Tech Gesellschaft für funktionelle Membranen und Anlagentechnologie mbH (GERMANY)

28

FLORIDA SOLAR ENERGY CENTER®
Creating Energy Independence

An Appreciation of Solid Hydrogen Storage and Release

Allison M. Fisher, Ph.D.
Visiting Professor – Consultant
Florida Solar Energy Center

1679 CLEARLAKE ROAD, COCOA, FL 32922-5703 • TEL: 321-638-1000 • FAX: 321-638-1010 • WWW.FLORIDAENERGYCENTER.ORG
A RESEARCH INSTITUTE OF THE UNIVERSITY OF CENTRAL FLORIDA • AN EQUAL OPPORTUNITY AND AFFIRMATIVE ACTION EMPLOYER

FLORIDA SOLAR ENERGY CENTER®
Creating Energy Independence

Portable hydrogen-fed fuel cell power systems won't be fully enabled until lightweight, compact hydrogen delivery systems are available **and** achieve >3x the energy density of the best battery technologies.

Some portable hydrogen-delivery cartridges are commercially available today:

- Early versions were based on intermetallic hydrides;
- Hydrolysis-based chemical hydrides followed;
- Newest systems are based on chemical hydrides.
- Pros: safe, easy to use
- Cons: packaged energy density is not >3x batteries

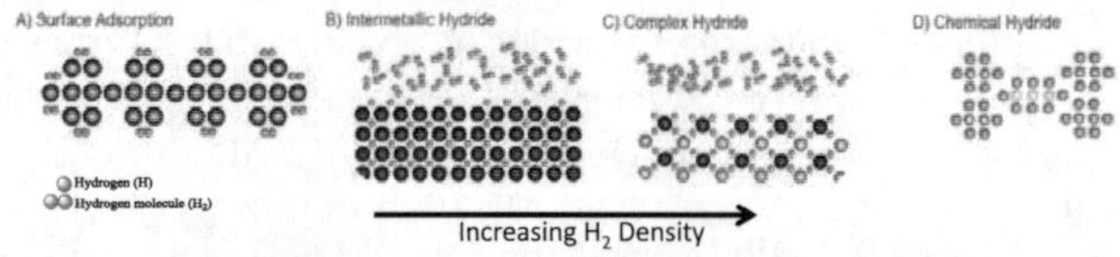

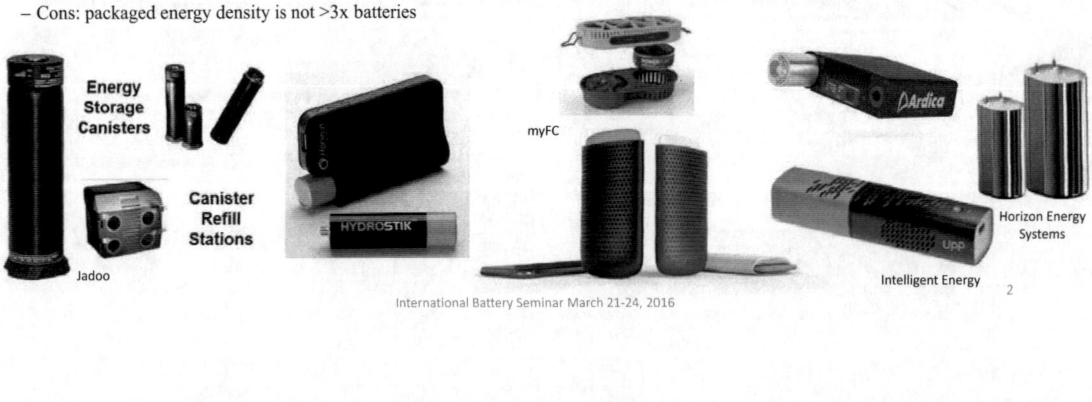

The fuels that have the best potential for achieving >3x battery energy density are chemical hydrides that can release hydrogen on heating.

FLORIDA SOLAR ENERGY CENTER®
Creating Energy Independence

- A primary hydrogen cartridge will satisfy the portable market if the energy density/specific energy is high enough.
- Cartridges based the thermolysis of solid chemical hydride fuels are not quite commercial yet:
  - ITAR* and related export control issues
  - Fuel availability and cost

  *ITAR = International Traffic in Arms Regulations

- What do these cartridges look like inside?
  - Solid fuel may be compartmentalized and packaged for controllable hydrogen release,
  - System controls are required to determine when to initiate hydrogen release depending on demand from the power system,
  - Few fuels are likely to release hydrogen with sufficient purity for a fuel cell system; consequently, a filtration unit will be required, and finally
  - Insulation will be needed to prevent one fuel compartment from setting off another ("cross-initiation").

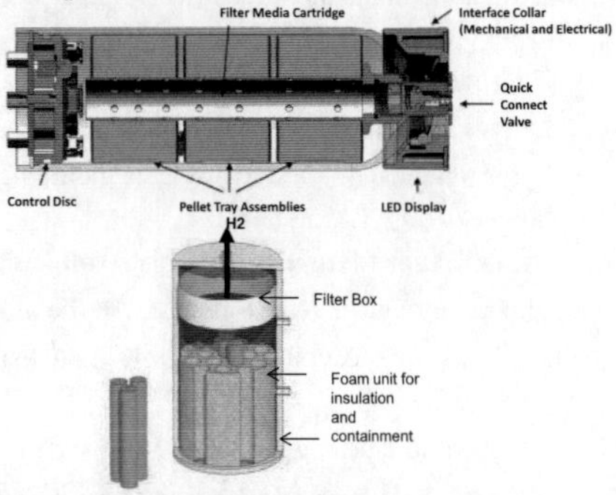

The more the internal system balance of plant, the lower the achievable W-hr/L; W-hr/kg.

In order to compete with batteries in the marketplace, a solid hydrogen cartridge must achieve >1000W·hr/L

FLORIDA SOLAR ENERGY CENTER®
Creating Energy Independence

- Fuel cell system efficiencies are NOT near 100%. A good system will operate at about 40% efficiency.
- It should be assumed that at least 40% of the cartridge space must be made available for BOP. Ideally 50-60% of the cartridge space will be available for fuel.

Alane (AlH_3) **Ammonia Borane** New Fuels

A cartridge of 1L volume must contain 75.1g releasable H2 for a fuel cell system operating at 40% efficiency.

Fuel Density (g/mL)	Fuel Hydrogen Capacity (wt%/100):															
	0.09	System/mL	0.10	System/mL	0.11	System/mL	0.12	System/mL	0.13	System/mL	0.14	System/mL	0.15	System/mL	0.16	System/mL
0.80	1043.1	-43.1	938.8	61.3	853.4	146.6	782.3	217.7	722.1	277.9	670.5	329.5	625.8	374.2	586.7	413.3
0.90	927.2	72.8	834.4	165.6	758.6	241.4	695.4	304.6	641.9	358.1	596.[illegible]	404.0	556.3	443.7	521.5	478.5
1.00	834.4	165.6	751.0	249.0	682.7	317.3	625.8	[illegible]74.2	577.7	422.3	536.4	463.6	500.7	499.3	469.4	530.6
1.10	758.6	241.4	682.7	317.3	620.7	379.3	568.9	43[illegible].1	525.2	474.8	487.7	512.3	455.2	544.8	426.7	573.3
1.20	695.4	304.6	625.8	374.2	568.9	431.1	521.5	478.5	481.4	518.6	447.0	553.0	417.2	582.8	391.1	608.9
1.30	641.9	358.1	577.7	4[illegible]2.3	525.2	474.8	481.4	518.6	444.4	555.6	412.6	587.4	385.1	614.9	361.1	638.9
1.40	596.0	404.0	536.4	463.[illegible]	487.7	512.3	447.0	553.0	412.6	587.4	383.2	616.8	357.6	642.4	335.3	664.7
1.50	556.3	443.7	500.7	499.3	455.2	544.8	417.2	582.8	385.1	614.9	357.6	642.4	333.8	666.2	312.9	687.1
1.60	521.5	478.5	469.4	530.6	426.7	573.3	391.1	608.9	361.1	638.9	335.3	664.7	312.9	687.1	293.4	706.6

One of the approaches to new fuels is to leverage the wealth of prior work on ammonia borane and overcome material challenges by blending.

FLORIDA SOLAR ENERGY CENTER
Creating Energy Independence

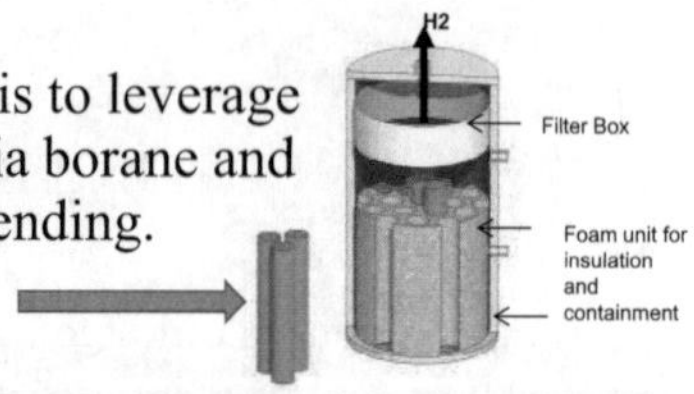

The challenges of ammonia borane as a hydrogen storage fuel include:

- Extracting all the theoretical wt% H_2: typically on 12-13wt% is readily available
- Hydrogen release is very exothermic
- Large volume expansion during dehydrogenation
- Large amounts of toxic byproducts including ammonia, aminoborane, borazine,
- Particulate that form from gas-phase oligomerization of aminoborane.
- The faster you heat AB to release H_2, the more impurities are generated
- AB is a strong reducing agent, will impact choice of blend material

As-received AB typically produces 11-12 wt% H_2 up to 170°C; purified AB will release 13 wt% H_2 under the same conditions.
pellet density ~0.78g/cm^3

Heating rate	Product yield (mol product per mole borazane)			
	H_2	BH_2NH_2	Borazine	Diborane
0.5 K min^{-1}	2.2	0.12	0.035	0.02
1.5 K min^{-1}	2.2	0.16	0.05	0.025
5 K min^{-1}	2.2	0.21	0.065	0.04
Iso 363 K–1 K min^{-1}	2.2	0.12	0.04	0.02

Baitalow, F.; *et al. Thermochim. Acta* **2002**, *391*, 159-168

GC/MS Analysis during dehydrogenation of AB shows ammonia, pentaborane, borazine, and higher B-N oligomers.

FLORIDA SOLAR ENERGY CENTER
Creating Energy Independence

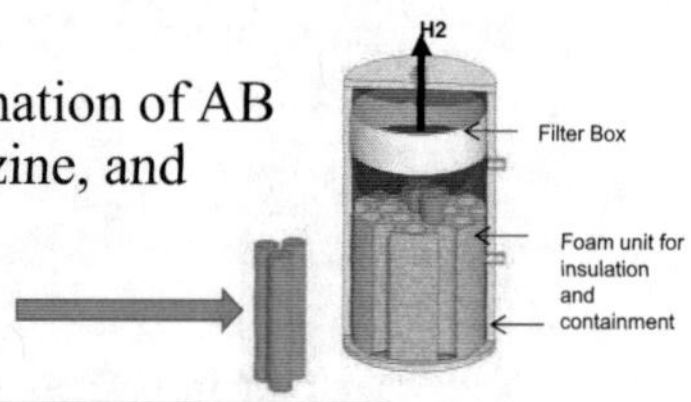

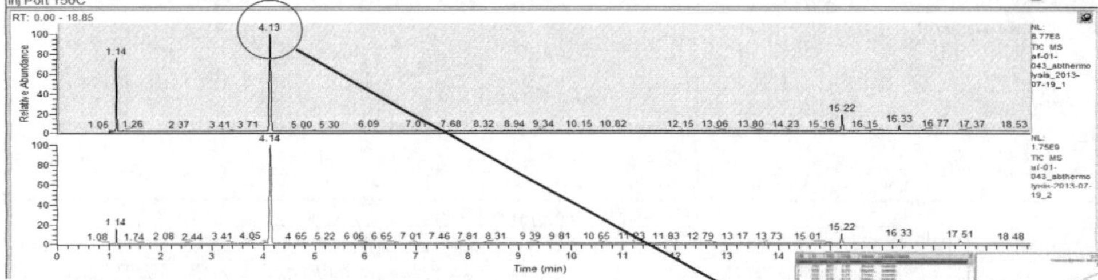

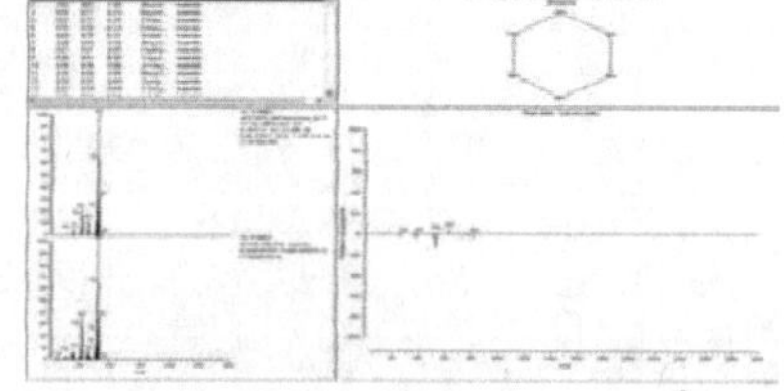

- Two aliquots were tested: one just as the hydrogen was released, and one after about 30 minutes of heating.

Ammonia was only observed in the first aliquot as soon as H_2 was released. Diborane was only observed as pentaborane.

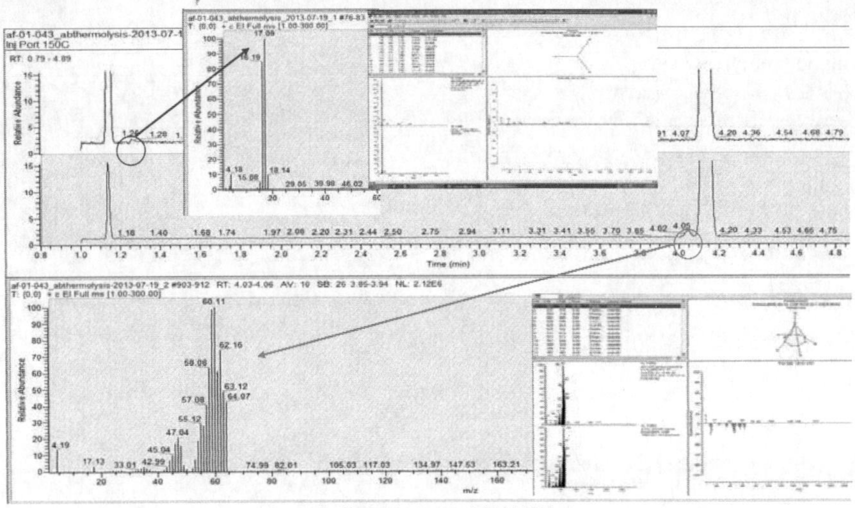

Ammonia was only observed in the first aliquot as soon as H_2 was released. Diborane was only observed as pentaborane.

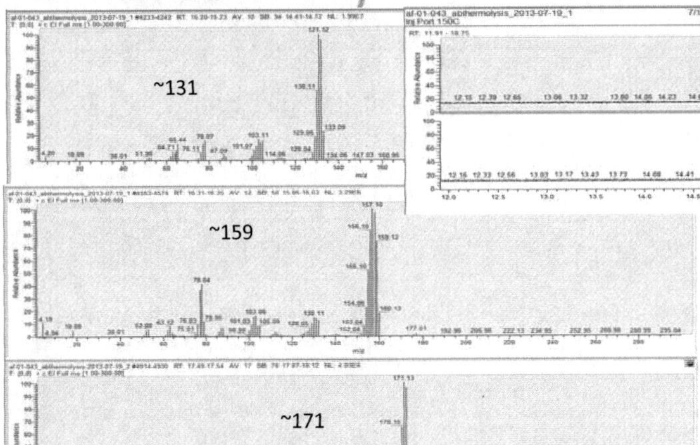

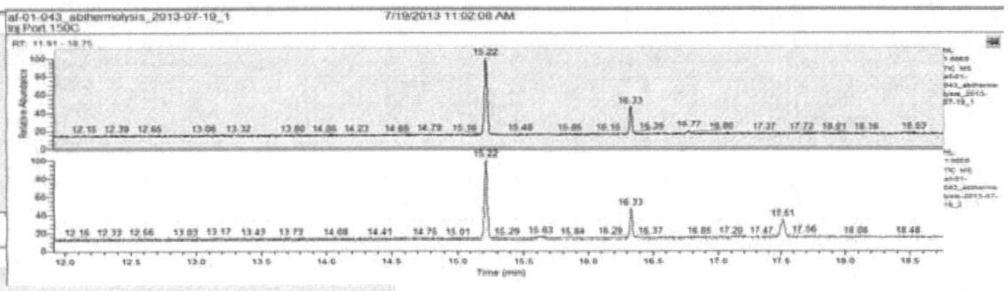

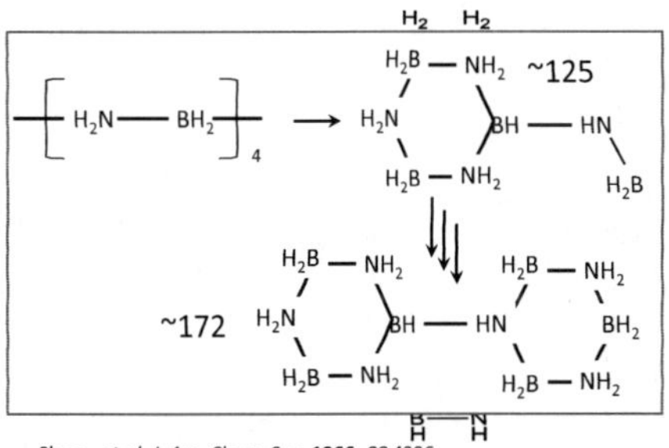

Shore, *et al. J. Am. Chem. Soc.* **1966**, *88* 4396

There are many reports on AB blends; we were able to test a number of them in our laboratory and compare to unblended AB.

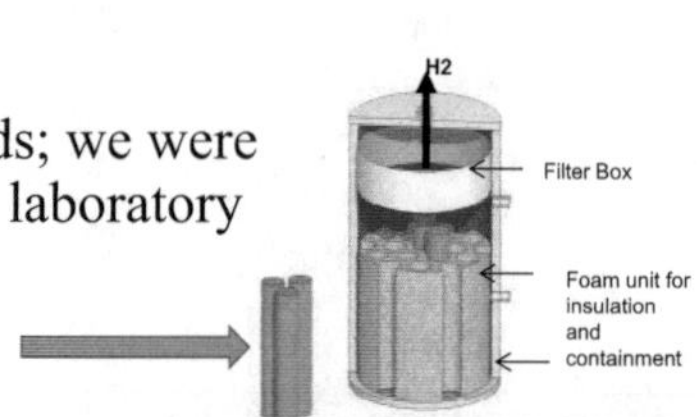

FLORIDA SOLAR ENERGY CENTER
Creating Energy Independence

Non-reactive polymers

- Blends with poly(ethyleneoxide):
  - Dehydrogenation less exothermic than AB
  - Smaller volume expansion during dehydrogenation
  - No clear reduction of gas-stream impurities compared to neat AB
- GC/MS results:

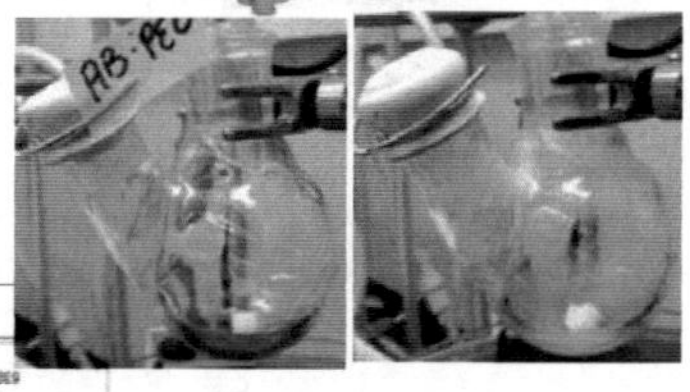

pellet before/after

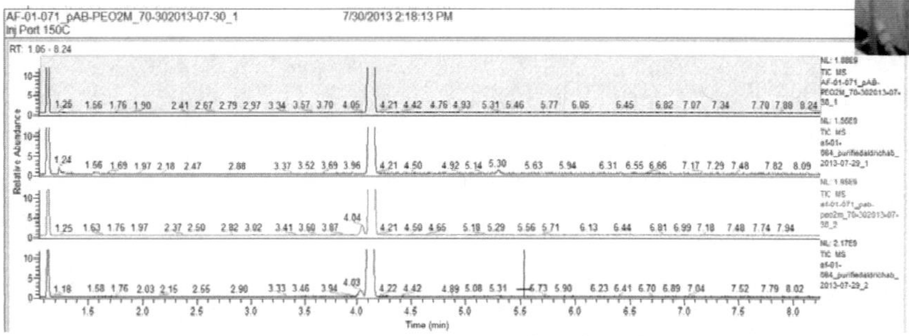

9 wt% H_2
pellet density 0.84g/cm^3

FLORIDA SOLAR ENERGY CENTER
Creating Energy Independence

Reactive polymers do affect the dehydrogenation of AB.

Reactive polymers

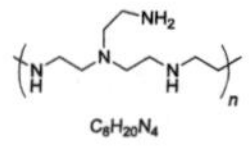

- Blends with branched polyethyleneimine (a liquid, viscous polymer):
  - Not able to do thermal analysis
  - Observed varying volume expansion during dehydrogenation
  - Some improvement in impurities at the expense of increased ammonia
- GC/MS results 80:20 and 70:30 AB:PEI blends:

9.9wt% H_2
not pelletized

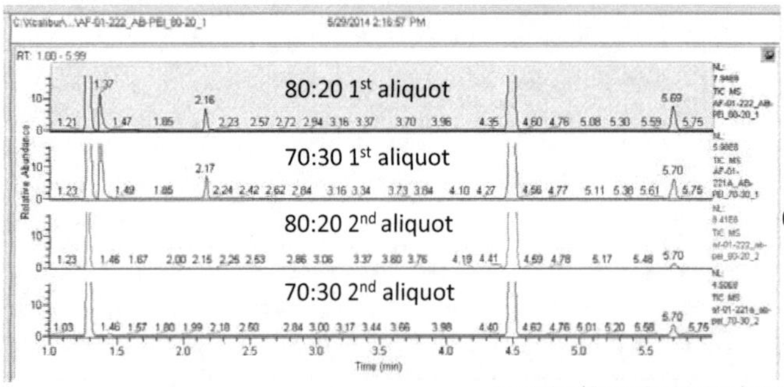

11wt% H_2
0.82g/cm^3

70:30 Blend

80:20 Blend

FLORIDA SOLAR ENERGY CENTER
Creating Energy Independence

Other things rarely described in the literature but which must be resolved before commercializing a fuel include:

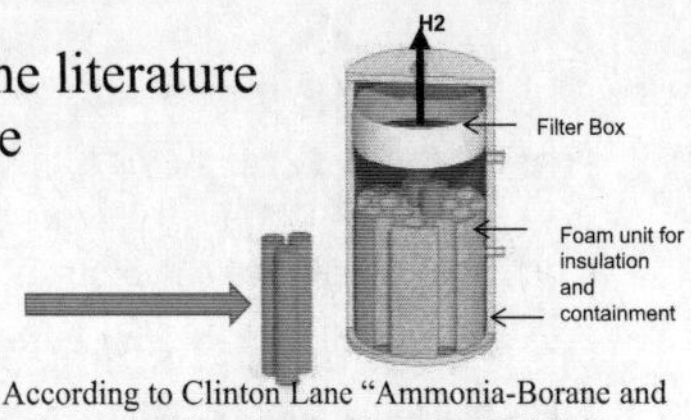

- Toxicity
  - AB is often touted as "safe" by some system developers
  - SDS suggests AB is not toxic; however detailed toxicological studies have not been reported. According to Clinton Lane "Ammonia-Borane and Related N-B-H Coumpounds and Materials: Safety Aspects, Properties and Applications" A survey completed as part of a project for the DOE CHSCofE, Contract #DE-FC36-05GO15060.
- Safety properties for transport; such as burn rate, self-heating, and self-accelerating decomposition temperature (SADT) (Anton, *et al. Int. J. Hydrogen Energy* **2012**, *37*, 2885)
  - AB shows no detectable reactivity with water
  - Burn rate of 33 mm/s; which is considered high and puts AB in a hazardous material class
  - AB failed the self-heating test: temp rose to 300°C within 10 min at an oven temp of 150°C, which is much higher than the allowable 60°C temperature rise.
- Lot-to-lot variability from manufacturer(s)
  - Significant differences in lot-lot purity from same manufacturer, as well as between manufacturers.
  - Even if differences are small, we have found that these levels impact mechanical and chemical properties on dehydrogenation, which in turn impacts the filter sub-system.
- Fuel shelf life:
  - Stability varied among manufacturers over the course of two years.
  - At worst, wt% H_2 dropped 1-2% and ammonia odor increased.
  - Need to determine effect of blending on shelf-life
- Effect of trace amounts of oxygen on dehydrogenation; or, limiting oxygen index.
  - Once a fuel is compressed it can be difficult to remove all the oxygen if the pellet is formed in air.

FLORIDA SOLAR ENERGY CENTER
Creating Energy Independence

A filtering sub-system is required to absorb impurities generated during thermal dehydrogenation.

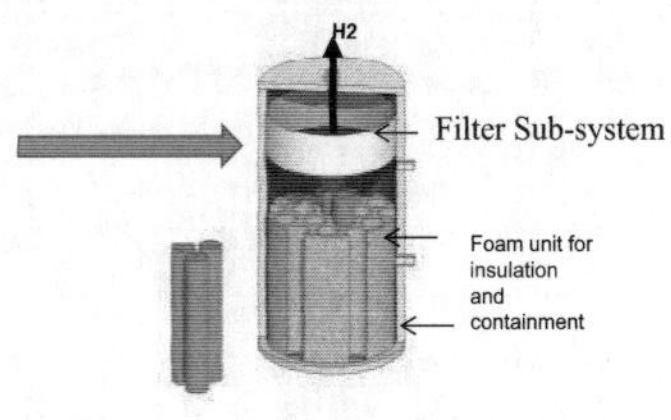

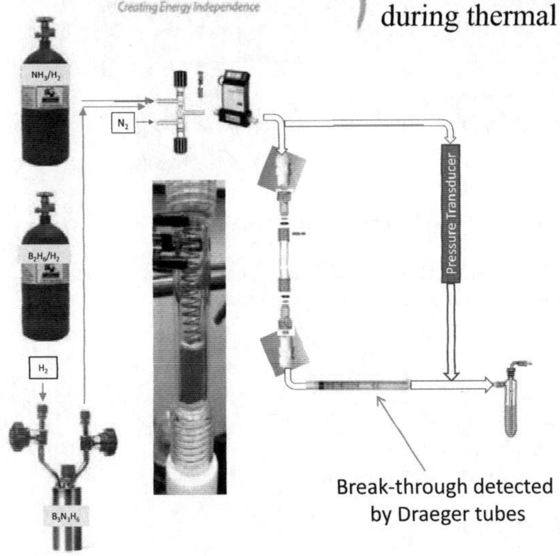

Break-through detected by Draeger tubes

- Few fuels will provide hydrogen at the purity level required of a fuel cell system
- We built a dynamic testing setup to screen sorbent materials
- Was capable of testing sorbents over -40°C to 80°C; up to 500sccm and 80psi

FLORIDA SOLAR ENERGY CENTER®
Creating Energy Independence

There are many materials available to screen for use as filter media.

- Criteria: 1) material availability; 2) cost; 3) pressure drop; 4) capacity; 5) selectivity
- Test setup benchmarked using sorbents reported by DOE for ammonia and results matched well.
- Since benchmarking the team tested >100 sorbent materials and retained a list of ~80 as potential candidates; additional custom-developed sorbents were identified and planned for next generation testing.
- *It was observed that sorbents that retained diborane and borazine without bonding chemically to them would ignite on contact with air during cleanup…!*
- Some of the better results with ammonia, diborane, and borazine are shown here:

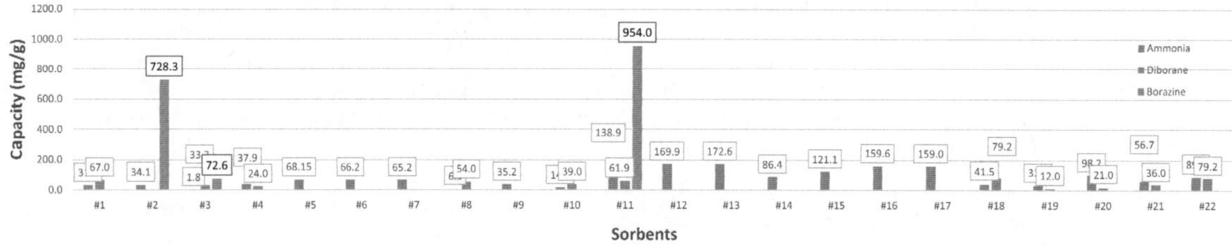

- For comparison: in similar work reported by the DoE, the best ammonia scubbing was on the order of 100-120mg/g; and borazine sorbent capacity was ca. 450mg/g.

FLORIDA SOLAR ENERGY CENTER®
Creating Energy Independence

The final area of the fuel system that involves chemistry is the insulation needed to prevent the fuel pellets from igniting other fuel pellets.

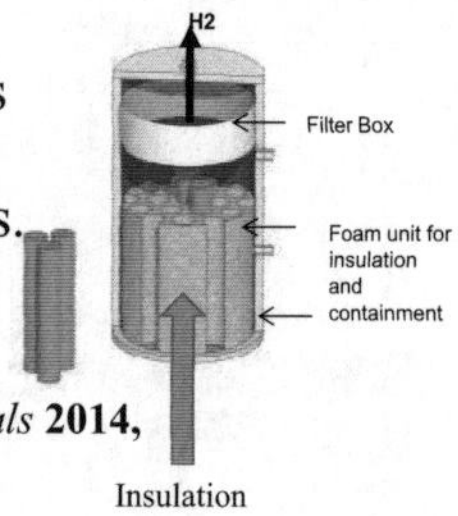

- Heat is transported through conduction and by convection in the hydrogen gas.
- Thermal characteristics of insulation material depend on (from Thapliyal, P.C.; Singh, K. *J. Materials* **2014,** 127049/1-127049/11):
  - Thermal conductivity of the material
  - Material cell matrix → most important
- Convection has a more significant impact. Since few materials are hermetic to hydrogen, the choices for preventing heat transfer to adjacent pellets are limited.
- The insulating materials have to withstand temperatures above 250°C, which also limits choices. In addition, care must be taken not to add impurities to the gas stream from the choice of insulation. Examples include:
  - Polyimide (Solimide)
  - Fiberglass
  - Mineral wool
  - Soda lime glass (FoamGlas™ One)
  - Silica, silica-alumina, ceramics
- The thicker the insulation layer; the less space is available in the cartridge for fuel.
- As a result, there isn't much available commercially that is optimal for compact, lightweight hydrogen fuel cartridges.
- Developments welcome in aerogels, others for use in solid hydrogen systems!

FLORIDA SOLAR ENERGY CENTER®
Creating Energy Independence

Concluding Remarks

- In order to commercialize a solid hydrogen storage system to power portable fuel-cell based devices, new fuels are needed that possess both higher gravimetric density and higher weight percent hydrogen.
- Other fuel properties that must be determined in order to commercialize a fuel include:
  - Thermal properties
  - Cost
  - Manufacturability
  - Burn rate, LOI, SADT tests
- Regarding AB, blends have potential for overcoming critical challenges, but the best blends have yet to be demonstrated.
- Filtration systems are essential subcomponents of a fuel cartridge; significant development remains in order to develop compact, lightweight units.
- The same is true for insulation around the fuel in order to prevent cross-initiation and thermal runaway.

International Battery Seminar March 21-24, 2016

15

FLORIDA SOLAR ENERGY CENTER®
Creating Energy Independence

Acknowledgements

- Cella Energy US (now closed):
  - Ms. Olivia P. Thomas
  - Dr. Jiezhu Jin
  - Ms. Nicole Carswell
  - Prof. Sean McGrady
- Cella Energy UK:
  - Dr. Arthur Lovell
  - Dr. Zeynep Kurban
  - Dr. Joseph Cook
- Florida Solar Energy Center (FSEC): Florida's energy research institute — with a 35-year history of unique expertise, experience and infrastructure — leading research and development efforts to bring our vision of Energy Independence to fruition.
  - Dr. Jim Fenton
  - Dr. Ali Raissi
  - Dr. Nazim Muradov
  - Dr. Nan Qin
- FSEC's Advanced Energy Research Division (AERD) facilities include three fully equipped chemical laboratories comprising 4750 ft^2 and a 1300 ft^2 field facility. Each of these laboratories is dedicated to a variety of hydrogen and fuel cell R & D programs and activities.

International Battery Seminar March 21-24, 2016

16

Gaseous Fueled Hybrid and Fuel Cell Vehicles (Natural gas as a bridge to hydrogen)

Andrew Burke
University of California-Davis
Institute of Transportation Studies

International Battery Seminar
Fort Lauderdale , Florida
March 21-24, 2016

ITS UCDAVIS

Outline of the presentation

1. Introduction
2. Gaseous fuel storage
3. Gaseous fueled vehicle designs
4. The Toyota Mira-layout and simulations
5. CNG hybrid and fuel cell vehicle simulations
6. Cost analyses
7. CO2 emission comparisons
8. Natural gas-hydrogen bridge considerations

ITS UCDAVIS

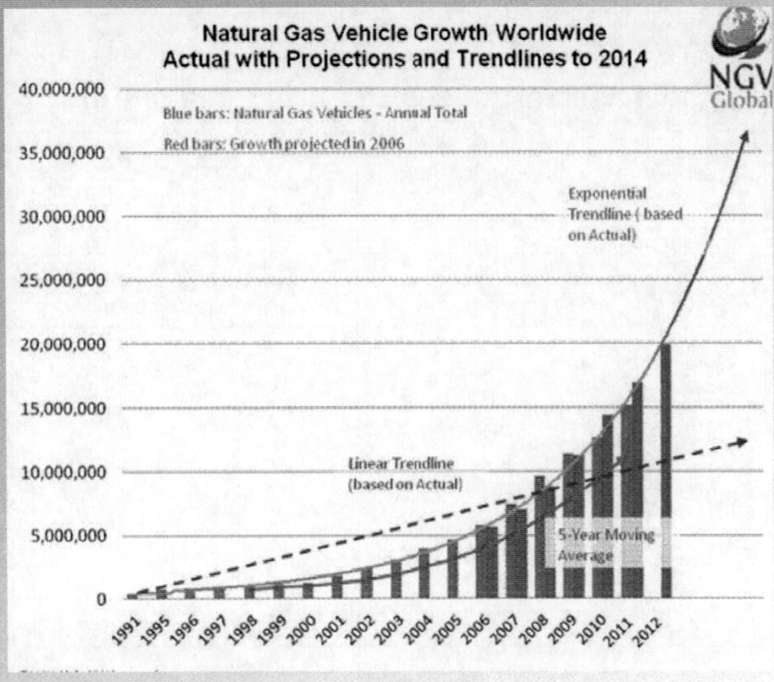

Source: NGV, Global

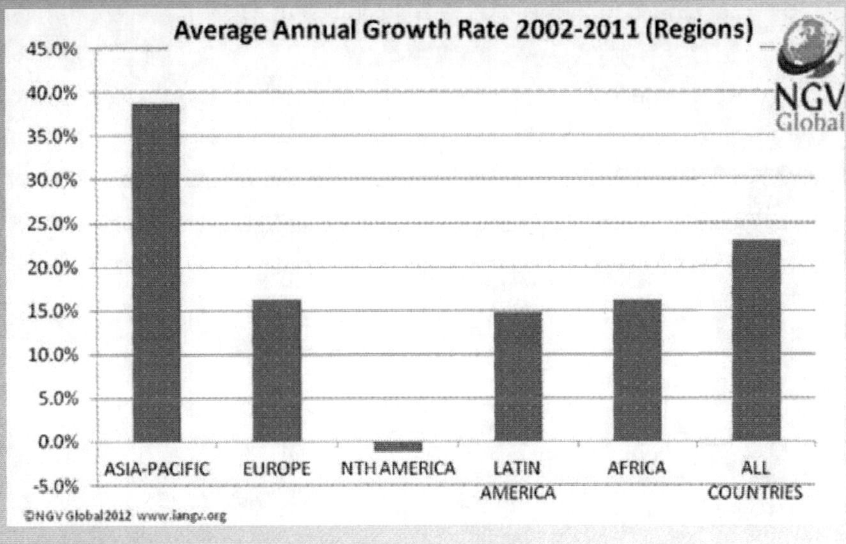

Source: NGV, Global

Bifuel natural gas vehicles being sold in Europe

Vehicle Type	Fuel Capacity		Range(combined)(Km[miles])	Engine Power(Kw[hp])	Emission Standard	CO2(g/Km)
	CNG (Kg)	Petrol(Kg)				
Fiat Panda	12	23	900[560]	60[80]	Euro5	86
Fiat 500 L	13	20	770[480]	63[85]	Euro5	98
Fiat Punto	13	30	1000[620]	52[70]	Euro5	115
Fiat Qubo	13	30	1000[620]	52[70]	Euro5	114
Fiat Fiorino	13	30	960[600]	52[70]	Euro5	119
Fiat Doblo	16	15	625[390]	90[120]	Euro5	134
Lancia Ypsilon Ecochic Methane	12	27	950[590]	60[80]	Euro5	86
Mercedes-Benz E200 NGT	19.5	36	1070[670]	122[163]	Euro5	149
Opel Zafira Tourer Turbo Eco	25	9	680[425]	112[150]	Euro5	129
Opel Combo	16-22	15	750[470]	90[120]	Euro5	134
Seat Mii	11	7	600[380]	51[68]	Euro5	79/86
Skoda Citigo	11	7	600[380]	51[68]	Euro5	79/86
VW Golf	18	34	1300[810]	82[110]	Euro5	99
VW UP	11	7	600[380]	51[68]	Euro5	79/86
VW Passat	21	21	940[590]	112[150]	Euro5	117
VW Touran	18-21	7	650[400]	112[150]	Euro5	128
VW Candy	37	9	760[480]	82[109]	Euro5	156
Audi A3	18	7	580[360]	82[110]	Euro5	99
Volvo V70	18	34	900[560]	159[213]	Euro5	164

Source: NGVA, Europe

HEV and FCV driveline schematics

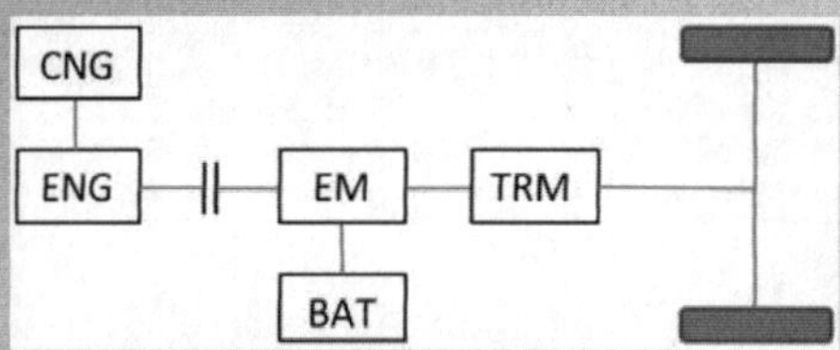

Natural gas hybrid

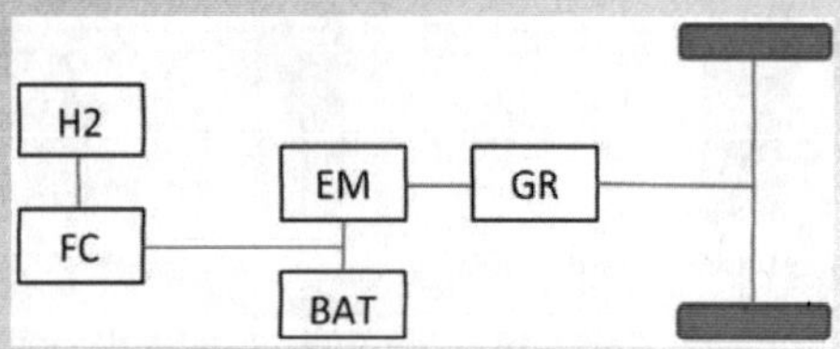

Hydrogen fuel cell vehicle

Energy Storage using natural gas and hydrogen

For Honda Civic GX

Energy Type	Volume (Gallon)	Weight(Kg)	Gallon Gasoline Equivalent (GGE)	Fuel Density (Kg/L)
Gasoline	8	21	8	0.72
CNG(3600 psi)	29	22	8	0.2
Hydrogen(5000 psi)	88	8	8	0.024
Hydrogen(10000 psi)	54	8	8	0.039

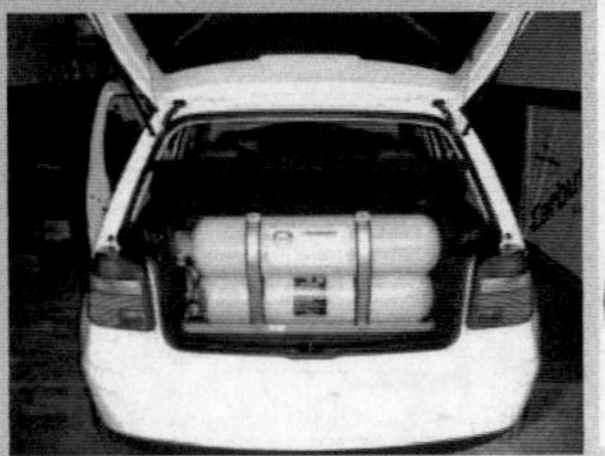

Honda Civic GX

Mercedes B 200 Dual fuel

More storage options are needed

Component layout for the Toyota Mira

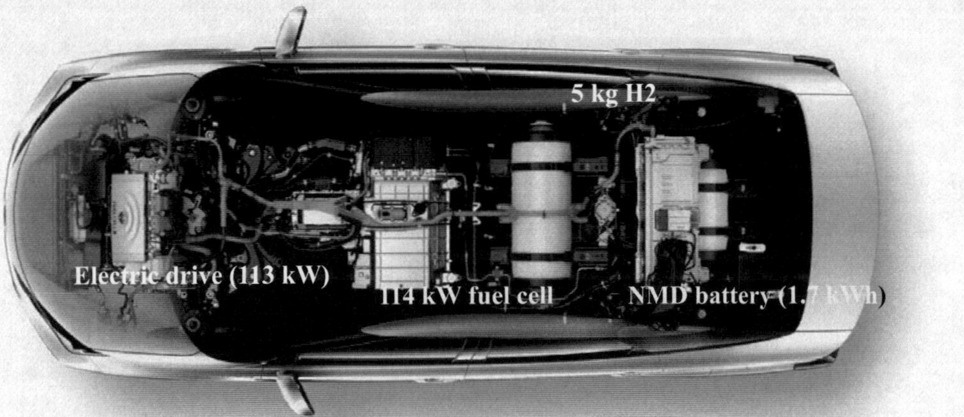

Inputs for the simulation of the Toyota Mira

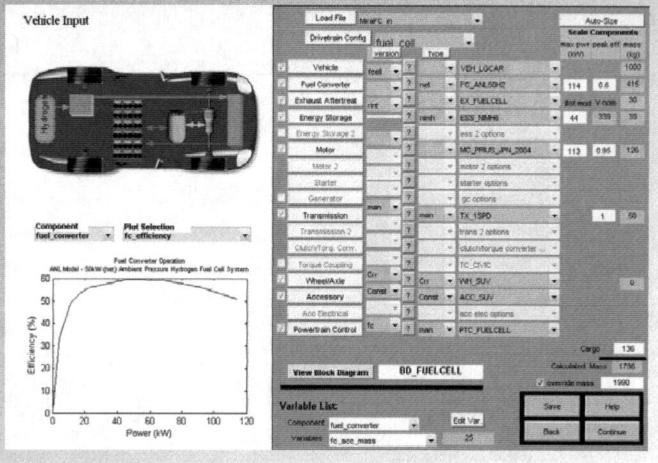

Nickel metal hydride battery 39 kg, 1.8 kWh

Carbon/carbon supercapacitor 21 kg 200Wh

Advisor simulations of the Toyota Mira with NMH batteries and supercapacitors

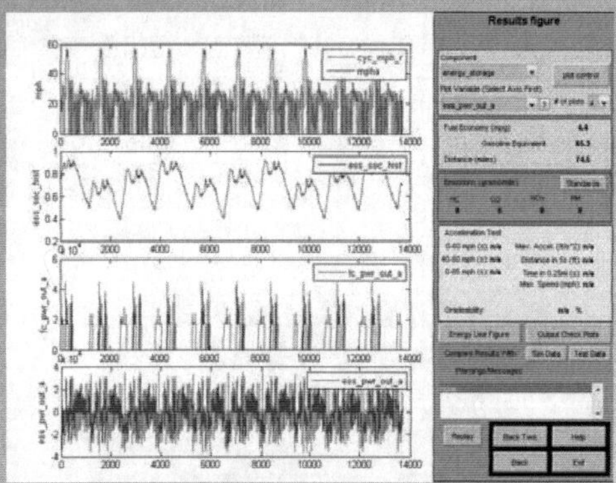

Toyota Mira FCV with nickel metal hydride batteries
Battery efficiency = 80%
Mpg-equiv. on FUDS 65.3

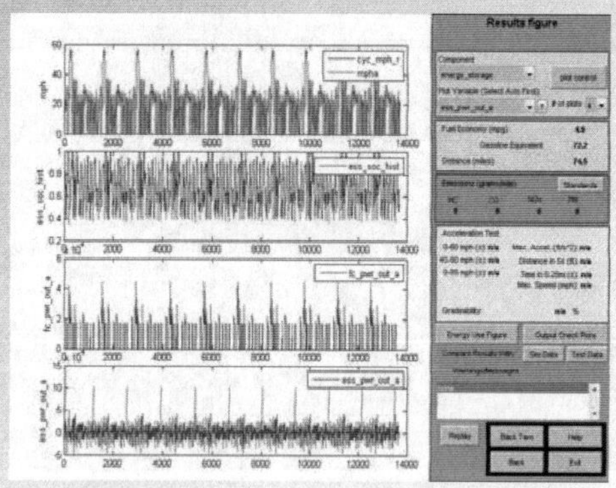

Toyota Mira FCV with carbon/carbon supercapacitors
Battery efficiency = 95%
Mpg-equiv. on FUDS 72.2

EPA fuel economy tests of fuel cell vehicles

Fuel cell Vehicle	Model year	City mpg*	Highway mpg*	Range miles
Toyota Mira	2016	66	66	312
Honda Clarity	2014	58	60	231
Hyundai Tucson	2016	49	51	265
	* 1 kg	H2 =	1 gal gasoline	

ITS UC DAVIS

Comparisons of energy storage for natural gas and hydrogen

Summary of gaseous fuel energy storage characteristics

Source or requirement	fuel	Gasoline equivalent Gal.	Tank volume (1) L	Tank weight (1) kg	MJ/L_{tank}	MJ/kg_{tank}
Present status	H_2 (2)	5 (1)	230	130	2.6	4.6
DOE goal 2015 (new)	H_2	5	128	130	4.7	6.5
Available composite	CNG (3)	5	80	33	7.5	18.2
Available metal	CNG	5	69	62	9.6	9.3

(1) 600 MJ fuel energy stored

(2) Hydrogen stored at 10000 psi

(3) CNG stored at 3600 psi

ITS UC DAVIS

Vehicle simulations with ADVISOR

Input parameters for the ADVISOR vehicle simulations

Vehicle type	C_DA m^2	f_r	Weight kg	Engine kW	Electric motor kW	Battery kWh	Fuel cell kW	Electric motor kW	Battery kWh
Compact	.6	.008	1388	97	15	1.0	60	95	1.0
Mid-size	.68	.008	1617	125	25	1.5	75	110	1.5
Full-size	.71	.008	1890	160	50	2.0	100	140	2.5
Small SUV	.72	.008	1700	140	25	1.5	85	120	1.5
Mid-size SUV	.75	.008	2100	150	40	2.0	100	125	2.0
Delivery truck	4.7	.008	7430	200	75	3.0	125	200	4.0

ITS UC DAVIS

Simulation results for various types of vehicles

Comparisons of the Storage requirements (Liters) for CNG in an HEV and H_2 in a fuel cell vehicle

		CNG HEV (1)			H_2 FCV			
Vehicle type	EPA Mpg ICE Vehicle	City mpg Gasoline Equiv.	HW mpg Gasoline Equiv.	CNG storage for 400 miles L (3)	City mpg Gasoline Equiv.	HW mpg Gasoline Equiv.	H2 storage for 400 miles L(3)	Ratio of mpg FC to HEV
Compact car	28/39	52.8	59.3	106	75	88	224	1.46
Mid-size car	27/36	49.4	53.1	116	71	79	244	1.47
Full-size car	25/36	44.3	51.3	124	67	75	258	1.48
Small SUV	22/30	47.3	51.9	120	69	80	246	1.50
Mid-size SUV	17/24	40	46	158	62	70	277	1.50
UPS delivery truck		12.0	11.3	128 L for 100 mile range	19	16.4	258	1.52

(1) CNG at 3600 psi, 8 MJ/L (steel tank)

(2) H_2 at 10,000psi, present technology 2.6 MJ/L

(3) Gasoline energy content 119 MJ/gal.

ITS UC DAVIS

Vehicle price projections 2015-2030

Assumed component costs in the cost analysis

Component	2015	2020	2025	2030
Fuel cell system $/kW	70	60	50	45
Electric motor/elect. $/kW	45	35	30	27
Lithium battery $/kWh	600	450	400	375
H_2 storage $/kWh	20	15	12	10
CNG storage $/kWh	9	7	6	5
Engine/trans. $/kW	42	42	42	42

Results of the price calculations for the CNG HEV and H_2 FCV

Vehicle type	Baseline Vehicle price $	Vehicle price W/o driveline $		2015	2020	2025	2030
Compact	19000	14800					
CNG HEV				26514	25231	24646	24161
H_2 FCV				33275	29510	27540	25690
Mid-size	25000	19000					
CNG HEV				33805	32215	31495	30925
H_2 FCV				40900	36490	34190	31985
Full-size	31000	23500					
CNG HEV				42603	40548	39595	38868
H_2 FCV				48775	43790	41200	39550
Small SUV	25000	19000					
CNG HEV				35605	33813	33010	32377
H_2 FCV				43165	38305	35817	33393
Mid-size SUV	33000	24750					
CNG HEV				42570	40508	39570	38847
H_2 FCV				52418	46860	44043	41265

Cost projections for natural gas and hydrogen vehicles

Projected prices of fuel cell powered mid-size cars

Year	Present study	Ogden, etc.
2010		49K
2015	41K	43K
2020	37K	39K
2025	34K	35K
2030	32K	31K

Price (2014 $) of mid-size passenger cars using natural gas and hydrogen

Year	$ Price baseline vehicle	$ Price Natural gas HEV	$ Price Hydrogen FCV
2015	25000	33805	40900
2020		32215	36490
2025		31495	34190
2030		30925	31985

CO2 emissions comparisons

Fuel type	Mi/kWh nat. gas	Kg CO2/mi
Electricity EV	1.44	.146
Natural gas HEV	1.39	.151
Hydrogen FCV	1.14	.185

Natural gas to electricity 40%

Natural gas to hydrogen 70%

Compressing natural gas 2.3%, compressing H2 7.2%

EV 250 Wh/mi of electricity from the wall plug

CNG HEV 50 mpg gasoline equivalent

H_2 FCV 75 mpg gasoline equivalent

1 kWh nat.gas = .0766 kg nat. gas

1 kg nat.gas = 2.75 kg CO2

Natural gas-hydrogen bridge considerations

- CNG sales are increasing worldwide
- CNG storage is more favorable than hydrogen storage
- CNG tanks for hybrids can be smaller than hydrogen tanks for fuel cell vehicles for the same range (400 miles)
- Vehicles can be designed for gaseous fuels without loss of utility to driver
- The cost of fuel cell vehicles will approach that of the CNG vehicles as the cost of fuel cells decreases
- CNG infrastructure could be used as bridge to providing hydrogen fueling stations

- <u>Thus a bridge to hydrogen fuel cells</u>

33rd International Battery Seminar: Fuel Cells Symposium

Fuel Cells for Unmanned Systems Propulsion

Jim Sisco, Paul Osenar

March 24, 2016

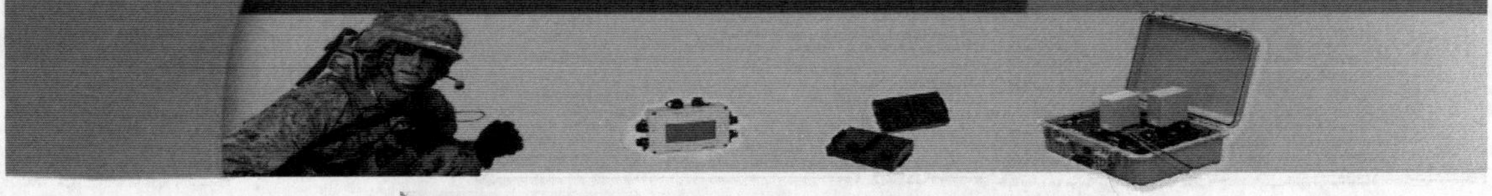

Protonex®

a Ballard® company

WWW.BALLARD.COM

WWW.PROTONEX.COM

Outline

Protonex
a Ballard® company

- Overview of Protonex Experience
- PEM Fuel Cell Stack Technology
- Sodium Borohydride (SBH) Fueling Solutions
- Power Management Systems
- PEM Fuel Cell Power Systems
- Unmanned Aerial Systems Applications
- Unmanned Undersea Power System
- Advanced Battery Charger
- Summary

Protonex Unmanned Systems Experience

Protonex a Ballard® company

- **Protonex PEMFC systems have strong unmanned system heritage:**
  - Decade of development
  - Implementation on multiple platforms and customers
  - 100's of hours of accumulated flight time
  - 3 active UAV power system development projects

- **Leverage several key Protonex technologies:**
  - Power dense PEM fuel cell stacks
  - High utilization (>99%) H_2 recapture schemes
  - Energy dense SBH fuel cartridges
  - Efficient power management electronics

NRL Ion Tiger UAV

Lockheed Martin Desert Hawk EER

QinetiQ Talon

PEM Fuel Cell Technology

Protonex a Ballard® company

- **Power dense PEM fuel cell stacks**
  - Adhesively bonded construction
  - Rugged and highly durable
  - Liquid cooled, closed cathode design
  - MEA agnostic

- **Specialized stack design for UAV systems**
  - Compact and lightweight
  - Demonstrated designs up to 1.5 kW
  - Up to 100 kW with Ballard technology
  - Specific power > 1400 W/kg

550 W Stack

1.5 kW Stack

Sodium Borohydride Fueling Solutions

Protonex
a Ballard® company

- **Safe, low cost chemical hydride**
  - Non-flammable
  - Non-toxic
- **High storage metrics**
  - Aqueous solution, SG = 1.0
  - 0.045-0.064 gH_2/g Solution
- **Hydrogen as needed**
  - Fast start-up
  - Rapid load following
  - Reliable control

Typical SBH System Flow Diagram

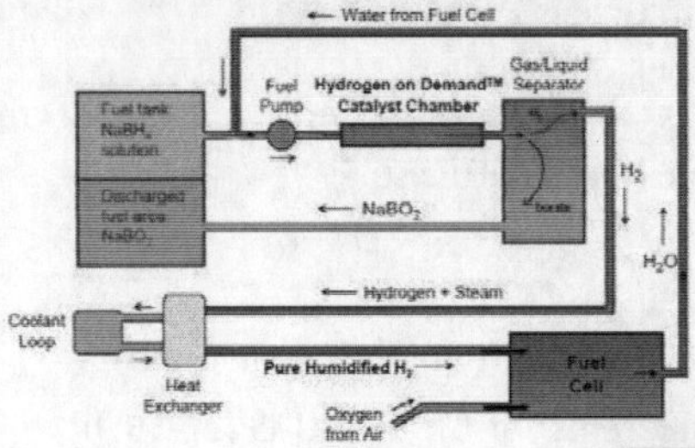

1350 Wh SBH Cartridges

Power Management Systems

Protonex
a Ballard® company

- **Efficient power management electronics**
  - Leverages Protonex power management products
  - Minimizes mission cost, maximizes endurance
  - Applicable to UAV platform and ground station power management
- **Enables hybridization schemes**
  - Prioritizes energy sources and sinks
  - Controls battery charge
- **Plug-and-play energy sources**
- **Intelligently directed power output**
  - Load shedding based on energy availability

Squad Power Management System

Typical UAV Power Management System

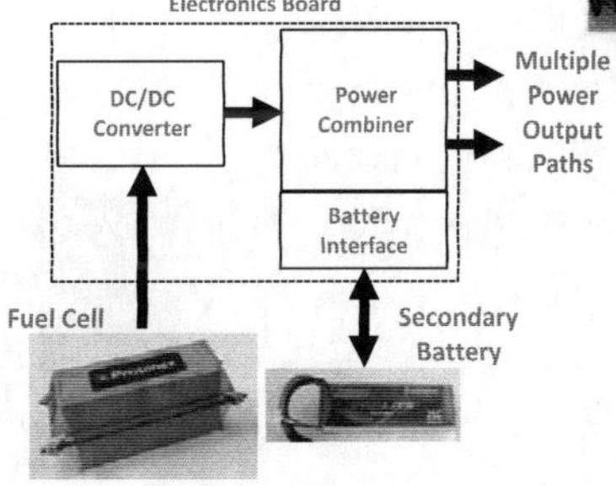

Complete PEM Fuel Cell Power Systems

Protonex a Ballard® company

- **SBH systems targeted to Group 1 UAS (<20 lb)**
  - Provide 2-3x specific energy of LiPo battery
  - Single-use fuel cartridges
  - Hybridized with LiPo battery for power bursts

SBH UAV Power System
350 W w/ Peaking Capability

- **Compressed/liquid H_2 systems for Group 2 UAS (21-55 lb)**
  - Low acoustic signature relative to IC engines
  - Exceptional throttle control
  - Reduced maintenance

550 W Compressed H_2 UAV Power System

Unmanned Aerial Systems Applications

Protonex a Ballard® company

- **550 W UAV Power System**
  - Designed for compressed/liquid H_2 systems
  - Demonstrated >99% H_2 utilization
  - Total mass = 1 kg (fuel cell, BOP, and electronics)

NRL Ion Tiger UAV

- **Powers Naval Research Lab (NRL) Ion Tiger platform**
  - 1100 Wh/kg and 24 hr endurance on compressed H_2
  - 48 hr endurance on liquid H_2

- **Powers NRL Sea Robin/XFC platform**
  - First submarine launched UAV

NRL Sea Robin/XFC UAV

Unmanned Undersea Systems Applications

Protonex a Ballard® company

- Air independent system leveraging hydrogen peroxide (H_2O_2) technology
  - Long heritage in rocket propulsion systems
  - Non-flammable, non-toxic
  - High storage metrics
    - Aqueous, SG = 1.2
    - 0.3 gO_2/g Solution
    - Oxygen on demand
- Protonex PEM stacks designed for pure O_2
  - Tailored flow passages
- 1 kW Benchtop demonstrator
  - 8 hr continuous operation w/ SBH
  - >99% O_2 utilization
- System estimated to provide >350 Wh/L
  - 2-3x increase over existing systems

1.5 kW O_2 Stack

UUV Power System Configuration

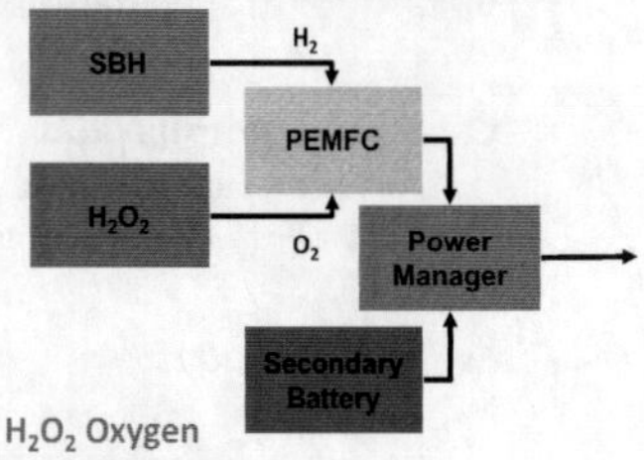

Benchtop H_2O_2 Oxygen Generation System

Advanced Battery Charger

Protonex a Ballard® company

- Quickly and efficiently charges up to 8 battery packs
  - Twice as fast and twice as efficient as common chargers
  - Standard modules for BB-2590, BB-2557, Li-80, and Li-145
  - Customized modules enable adaptation for <u>any battery design</u>
  - Charges UAV platform LiPo and ground station batteries
- Powered via AC, vehicles, solar, or generator energy
  - Worldwide AC adaptability
- Intelligent charging algorithm
  - Evaluates sources of power
  - Adjusts charge rate
  - Prioritizes charging
- Rugged, lightweight, waterproof

ABC System

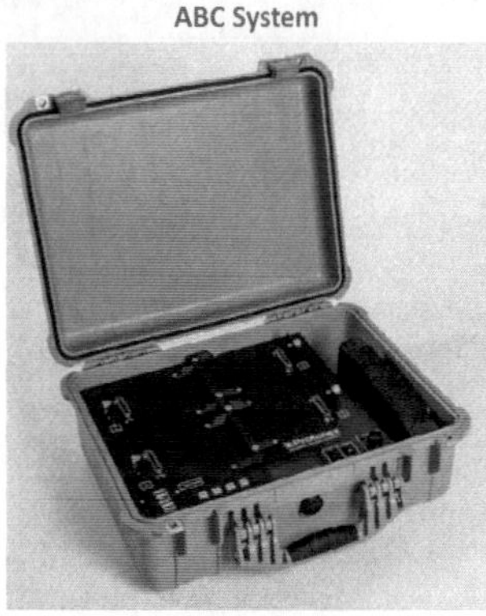

Battery Adapters

© 2016 Protonex Technology Corp. All Rights Reserved.

Summary

Protonex
a Ballard® company

- **Ballard / Protonex has over a decade of UxV experience**

- **Capability to design and manufacture fuel cell power systems for UxV from 100 W to 100 kW**
  - Propulsion, auxiliary power, sensors
  - Hand launched UAVs to commercial APUs

- **Expertise includes:**
  - Fuel cell stacks
  - Fuel cell systems
  - Fueling solutions
  - Battery selection
  - Power electronics

- **Continue to pursue new technologies and applications**

MeOH Power

Walter L. Robb
CEO
MeOH Power
874 Old Albany Shaker Road
Latham, NY 12110-1416
518-782-0050
waltrobb2@yahoo.com

Methanol – a great fuel

- **Safe -- Hard to Ignite -- Wood Alcohol**
- **Plentiful -- derived from any "organic" plus CO_2 + H_2**
- **Convenient -- melting point of - 98° C boiling point of + 65° C**

Methanol – a great fuel

- **Safe -- Hard to Ignite -- Wood Alcohol**
- **Plentiful -- derived from any "organic" plus CO_2 + H_2**
- **Convenient -- melting point of - 98° C boiling point of + 65° C**

Every gas station can readily install a Methanol Pump

Some History

MTI, Inc. licensed patents on methanol fuel cells from Los Alamos.

Company established MTI MicroFuel Cells, a subsidiary in 2001.

It concentrated on developing a 1-watt power device for cell phones.

MTI MicroFuel Cells switched to methanol vapor in its fuel cells to take advantage of the following characteristics

- ***fast diffusion in the anode chamber***
- ***no water required to mix with the fuel***
- ***no water-methanol loop***
- ***can start at*** *- 20° C*
- ***operates in absolutely dry air***

25+ patents on the "Mobion" technology

Historically, as you know,

Many companies have created methanol fuel cells

Most convert MeOH and H_2O to hydrogen and are really making H_2 fuel cells

Some make Direct Methanol Fuel Cells (DMFC):

They mix one mole of water with one mole of methanol and require a recirculating anode loop to add fuel and remove CO_2

Disadvantages: adds weight and complexity to the fuel and the system

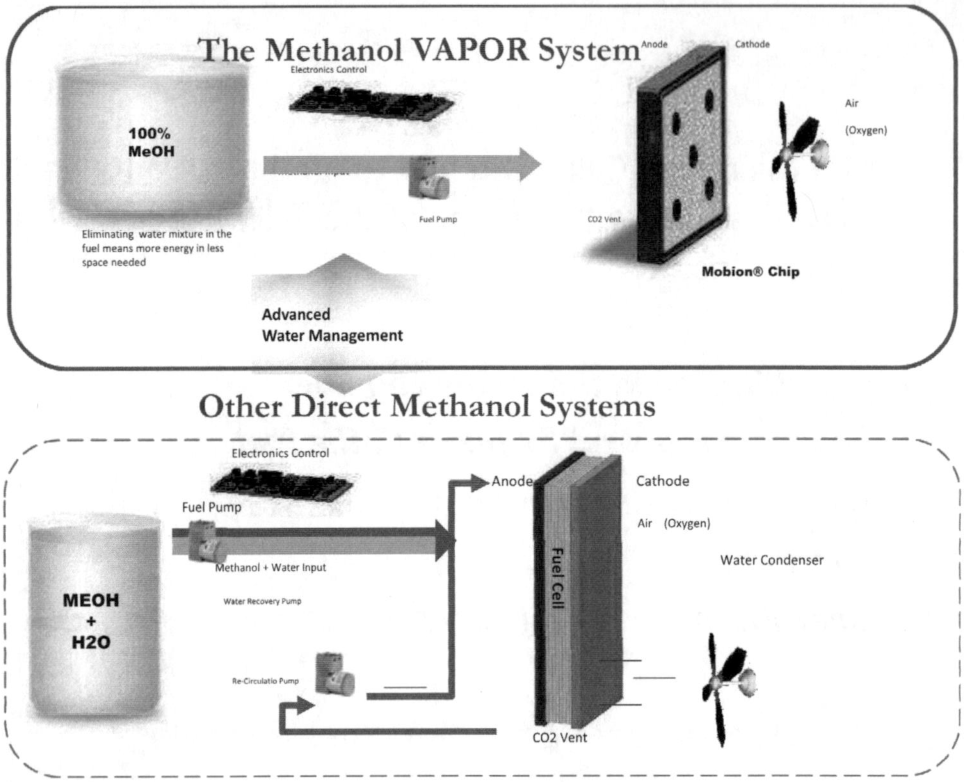

Cell phone batteries got smaller and cheaper

- MTI MicroFuel Cells made several hundred Mobion Rechargers
- U.S. Army proved the Rechargers were the only fuel cells to work in New Mexico field maneuvers
- But Lithium Ion batteries grabbed the cell phone market as well as the Recharger market.
- MTI MicroFuel Cells discontinued it efforts around 2012

Recent History

In 2014, the company reorganized:

MTI MicroFuel Cells became MeOH Power

- Emphasis on lightweight power devices
- Fuel weight – 0.8 gram of MeOH per "net watt hour," versus

  DMFC at 1.3 gram / wt. hr.

  and Li Ion battery at ~ 7 gram/ wt. hr.

Recent History

In 2014, the company reorganized:

MTI MicroFuel Cells became MeOH Power

- Emphasis on lightweight power device
- Fuel weight – 0.8 gram of MeOH per "net watt hour," versus

  DMFC at 1.3 gram / wt. hr.

  and Li Ion battery at ~ 7 gram/ wt. hr.

Of course, with a battery, but no FC is required

Starting with the Consumer Mobion System
and a design for a 100-watt fuel cell

Formula developed to estimate weight of a higher-powered system

with a 1-watt basic system weighing 92 grams

Actual Fuel Cell -- $\mathbf{W_{FG} = 18\ grams\ x\ P(watts)}$
[based on 55 milliwatts / cm^2]

For everything else -- $\mathbf{W_{auxiliaries} = 74\ grams\ x\ P^{0.6}\ (watts)}$
$W^{0.6}$ (watts)

thus, **(10)**$\mathbf{W_{FCS} = 180\ grams\ x\ 296\ watts = 476\ grams}$

a 10-Watt System is being built now:

5 layers of 50 cm^2 MEA's are key to the cell

MeOH Power has also scoped out several high watt cells:

Weight (wgt) = 18 grams x W(watts) + 74 grams x $W^{0.6}$ (watts)

Power	Total Weight	Fuel Weight for 5 hrs.
10 watts	1 lb / 0.46 kg	0.09 lbs / 40 grams
100 watts	7 lbs / 3.2 kg	0.9 lbs / 400 grams
1,000 watts	50 lbs / 22.0 kg	9 lbs / 4 kg
10,000 watts	440 lbs / 200 kg	99 lbs / 40 kg
20,000 watts	860 lbs / 390 kg	189 lbs / 80 kg

MeOH Power has also scoped out several high watt cells:

Weight (wgt) = 18 grams x W(watts) + 74 grams x $W^{0.6}$ (watts)

Power	Total Weight	Fuel Weight for 5 hrs.
10 watts	1 lb / 0.46 kg	0.09 lbs / 40 grams
100 watts	7 lbs / 3.2 kg	0.9 lbs / 400 grams
1,000 watts	50 lbs / 22.0 kg	9 lbs / 4 kg
10,000 watts	440 lbs / 200 kg	99 lbs / 40 kg
20,000 watts	860 lbs / 390 kg	189 lbs / 80 kg

Can MeOH Power do better?

The company has operated cells at 80 milliwatts/cm^2

That would reduce the 390 kg for 20,000 watts to 287 kg or 632 lbs. That would be very interesting for mobile power.

Issues

<u>Catalyst</u>

MeOH Power uses Pt/Ru at the anode
and PT at the cathode

For drones and devices requiring less than 1000 watts, the catalyst cost in not overwhelming

Issues

<u>Catalyst</u>

MeOH Power uses Pt/Ru at the anode
and PT at the cathode

For drones and devices requiring less than 1000 watts, the catalyst cost in not overwhelming

---

Yes, precious metals can be recycled.

MeOH Power's next goal is to greatly reduce the amount of such elements required in the cell.

FuelCell Energy
Ultra-Clean, Efficient, Reliable Power

Fuel Cell and Battery Hybrid for Energy Storage

Battery for Rapid Response and Fuel Cells for Range Extension for Storage

Pinakin Patel - Fort Lauderdale, March 24, 2016

Ultra-Clean | Efficient | Reliable Power

FuelCell Energy
Ultra-Clean, Efficient, Reliable Power

Integrated Fuel Cell Company

Research & Development

Design megawatt–class distributed power generation solutions

- *Global fuel cell platform*
- *Robust intellectual property portfolio*
- *Developing hybrid applications of existing technology for new markets*

Sales, Manufacture & Project Execution

Project development
- *Direct sales*

Global manufacturing profile
- *North America*
- *Asia via partner*
- *Europe*

Engineering, Procurement and Construction
- *Project Financing*

Services

Operate & Maintain power plants

- *Over 100 DFC® plants operating at more than 50 sites in 9 countries*
- *>3 billion kWh ultra-clean power produced*
- *> 300 MW installed/backlog*

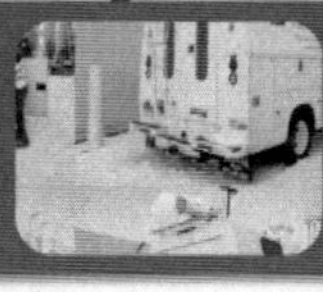

Providing turn-key distributed power generation solutions

NASDAQ: FCEL

2

FuelCell Energy
Ultra-Clean, Efficient, Reliable Power

Enabler for H_2 Infrastructure for Mfg., Transportation and Grid Support

Individual fuel cell
&
350 kW fuel cell stack

Four-Stack Module
1.4 megawatts

Completed module
1.4 megawatts

1.4 MW
DFC1500®
- *Utilizes one module*
- *Adequate to power 1,400 homes*

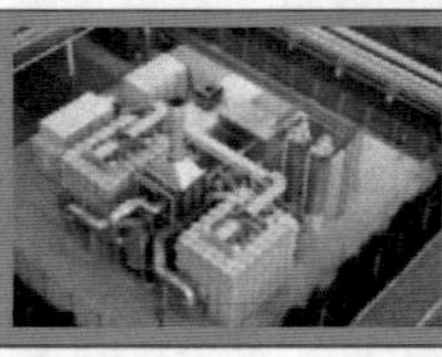

2.8 MW
DFC3000®
- *Utilizes two modules*
- *Adequate to power 2,800 homes*

59MW fuel cell park

DFC provides strong foundation for Tri-gen technology (70 MW/yr production rate)

3

FuelCell Energy
Ultra-Clean, Efficient, Reliable Power

Examples of Grid Support Needs

Storage Grid Domains (Grid Interconnection Point)	Regulatory Function	Use-Case Examples
Transmission-Connected	Generation/Market	**(Co-Located Energy Storage)** Concentrated Solar Power, Wind + Energy Storage, Gas Fired Generation + Thermal Energy Storage
		(Stand-Alone Energy Storage) Ancillary Services, Peaker, Load Following
	Transmission Reliability (FERC)	Voltage Support
Distribution-Connected	Distribution Reliability	Substation Energy Storage (Deferral)
	Generation/Market	Distributed Generation + Energy Storage
	Dual-Use (Reliability & Market)	Distributed Peaker
Behind-the-Meter	Customer-Sited Storage	Bill Mgt/Permanent Load Shifting, Power Quality, Electric Vehicle Charging

Needs include on-site generation to distribution and transmission at multi-MW level

4

FuelCell Energy
Ultra-Clean, Efficient, Reliable Power

Demand-Response Needs for CA

CPUC Energy Storage Procurement Targets

Storage Grid Domain Point of Interconnection	2014	2016	2018	2020	Total
Southern California Edison					
Transmission	50	65	85	110	310
Distribution	30	40	50	65	185
Customer	10	15	25	35	85
Subtotal SCE	**90**	**120**	**160**	**210**	**580**
Pacific Gas and Electric					
Transmission	50	65	85	110	310
Distribution	30	40	50	65	185
Customer	10	15	25	35	85
Subtotal PG&E	**90**	**120**	**160**	**210**	**580**
San Diego Gas & Electric					
Transmission	10	15	22	33	80
Distribution	7	10	15	23	55
Customer	3	5	8	14	30
Subtotal SDG&E	**20**	**30**	**45**	**70**	**165**
Total - all 3 utilities	**200**	**270**	**365**	**490**	**1,325**

1,325 MW needed by 2020 – minimum
→ Market Potential for Tri-gen: up to 500 MW

5

FuelCell Energy
Ultra-Clean, Efficient, Reliable Power

Grid Support Services Using Fuel Cells

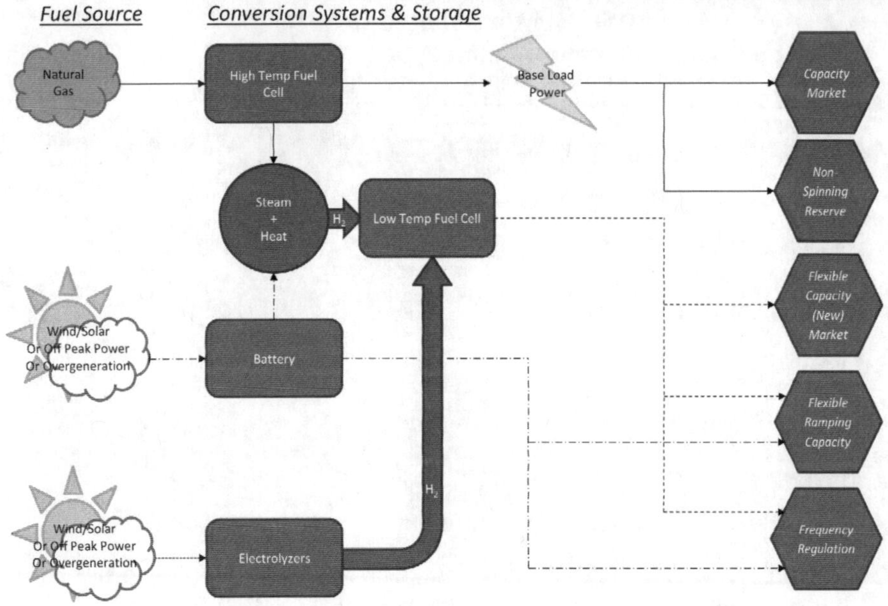

6

FuelCell Energy
Ultra-Clean, Efficient, Reliable Power

Grid Benefits: Tri-gen System

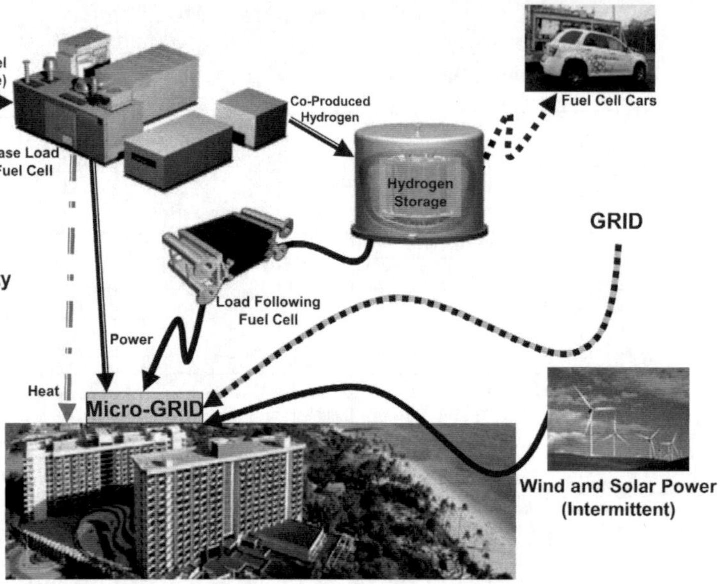

- Load Following
- Enhanced Energy Security
- Ultra Clean
- Provides Distributed H_2 for Black Start
- Supports Intermittent Renewables
- Fuel Flexible, Water Independent

Hydrogen storage for rapid demand response and black start

7

FuelCell Energy
Ultra-Clean, Efficient, Reliable Power

3Year+Testing Completed at OCSD

DFC-H2® Unit Started Up 9/15/10
at Orange County Sanitation District (OCSD) Fountain Valley, California

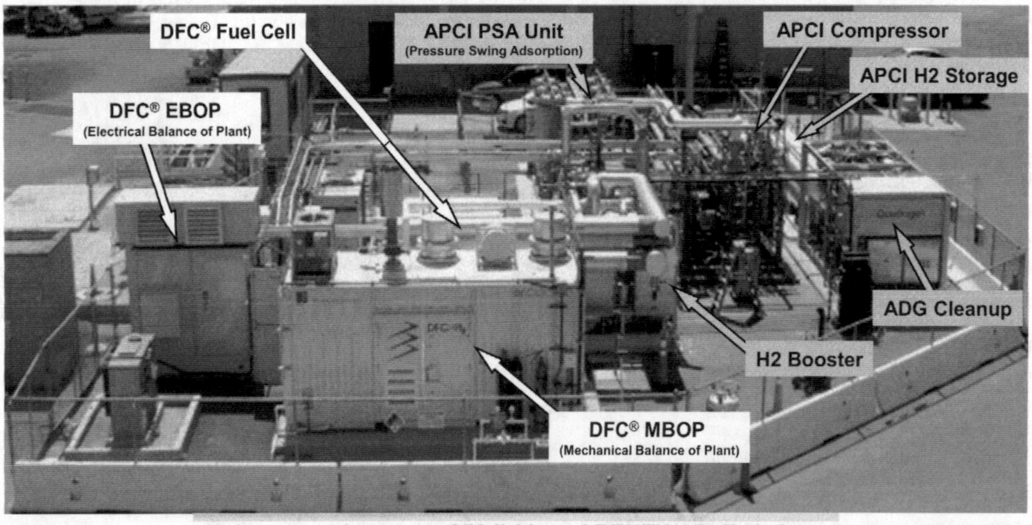

FuelCell Energy

Unit can produce over 250 lb/day of RENEWABLE H_2 from waste water anaerobic digester gas started in Jan 2011

AIR PRODUCTS

8

FuelCell Energy
Ultra-Clean, Efficient, Reliable Power

Industrial Application of Tri-Gen

Demonstrating the Value Proposition While Reducing Torrington Operating Costs

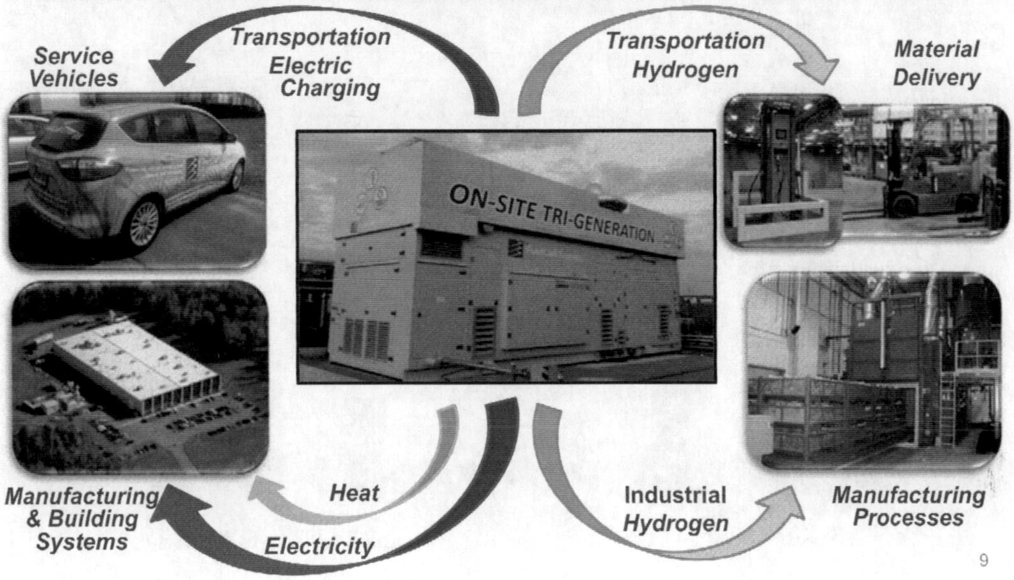

9

FuelCell Energy
Ultra-Clean, Efficient, Reliable Power

Demand Response Capacity of Tri-gen

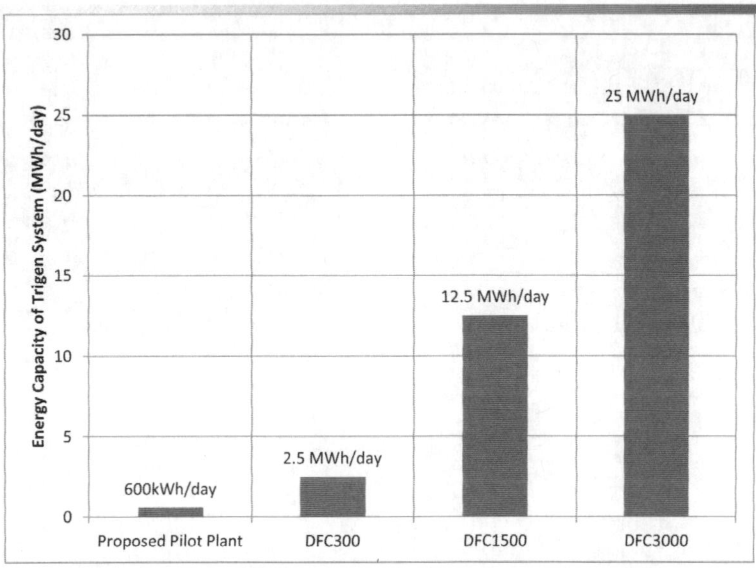

DFC-3000 provides up to 25 MWh/day

10

FuelCell Energy
Ultra-Clean, Efficient, Reliable Power

Advanced Energy Storage for Grid Support

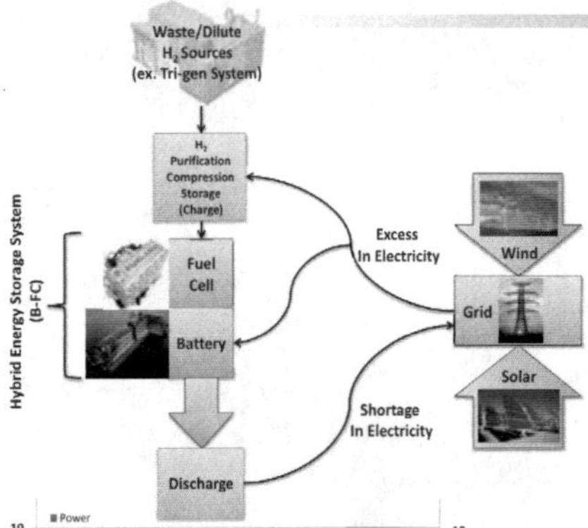

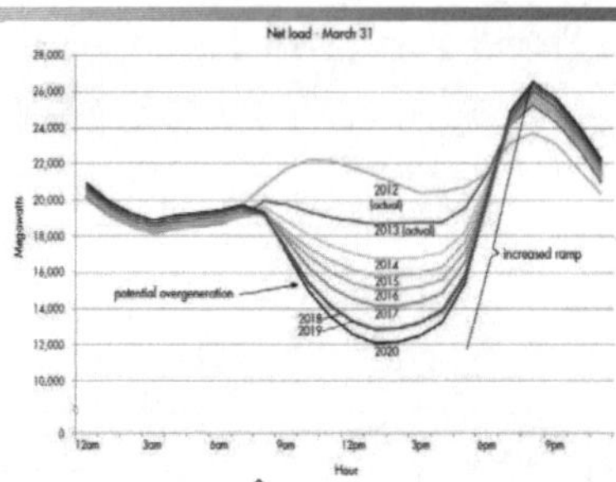

Solution to Duck Curve Issue

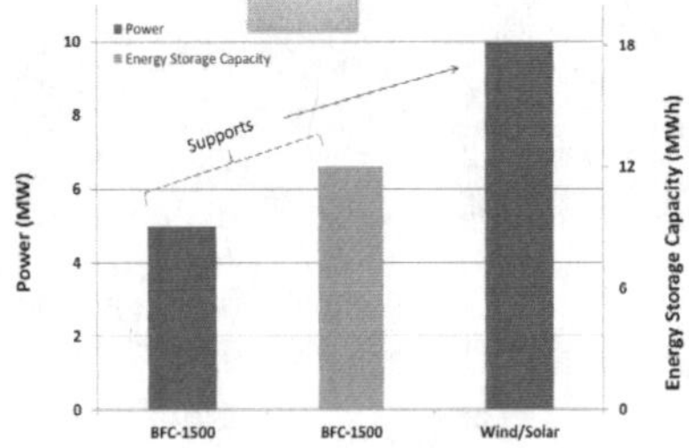

BFC-1500 System

x100 Systems

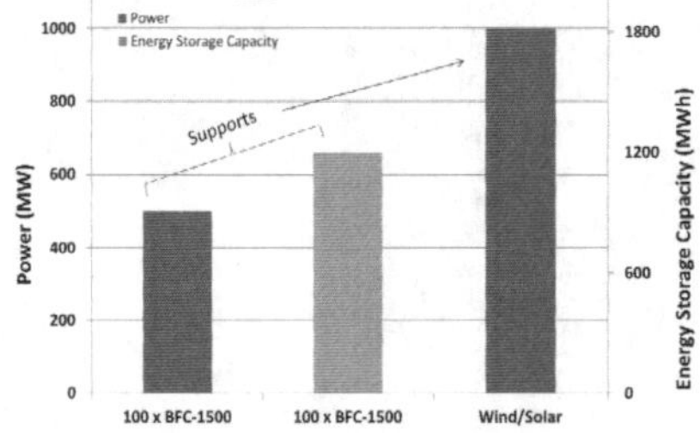

Cumulative Grid Support

11

FuelCell Energy
Ultra-Clean, Efficient, Reliable Power

Hydrogen for Energy Storage

14.9 MW fuel cell park Bridgeport, CT

59 MW fuel cell park Korea

	14.9 MW	59 MW
Hydrogen Produced	> 25 tons/day	> 100 tons/day
Hydrogen Available	~ 6 tons/day	~ 24 tons/day
Grid Support	**125 MWh**	**500 MWh**

Tri-gen system is enabler for demand response H_2

12

FuelCell Energy
Ultra-Clean, Efficient, Reliable Power

Reformer Electrolyzer Purifier (REP)

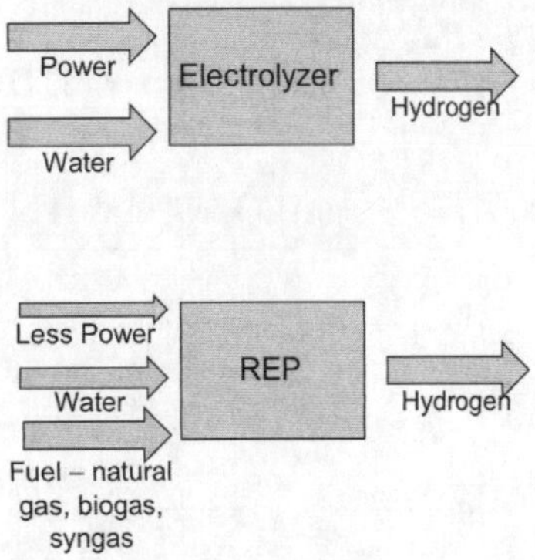

Advanced concept in early stage of development

- Reforms fuel while electrolyzing steam and purifying resulting hydrogen
- Combines H_2 production with purification – exploits CO_2 pumping of carbonate system
- Allows production of H_2 from variety of fuels and from syngas, but requires power

13

FuelCell Energy
Ultra-Clean, Efficient, Reliable Power

Standard Fuel Cell

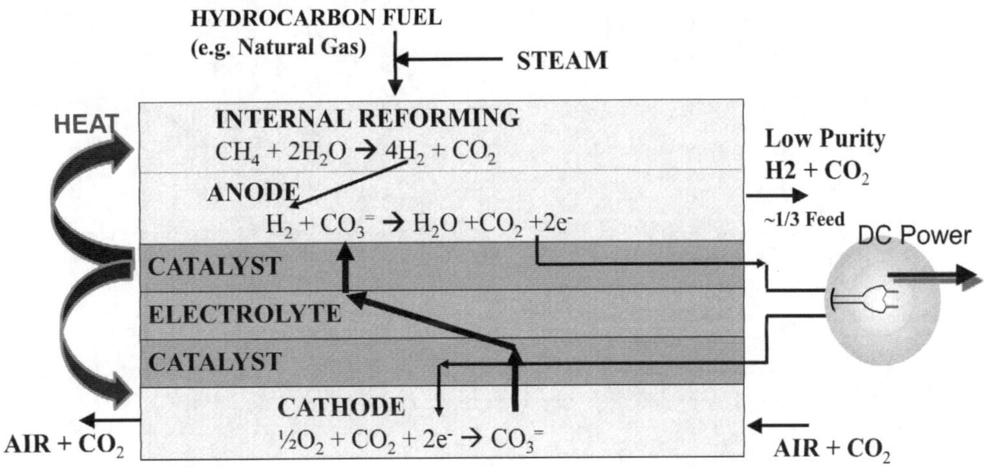

14

FuelCell Energy
Ultra-Clean, Efficient, Reliable Power

REP Cell

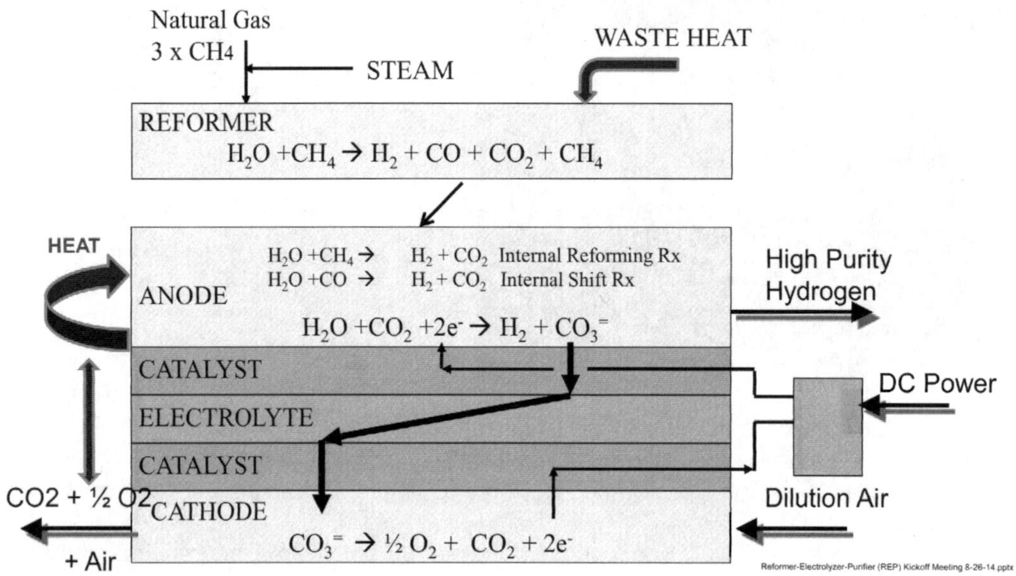

Reformer-Electrolyzer-Purifier (REP) Kickoff Meeting 8-26-14.pptx

15

FuelCell Energy
Ultra-Clean, Efficient, Reliable Power

Heat Level Required

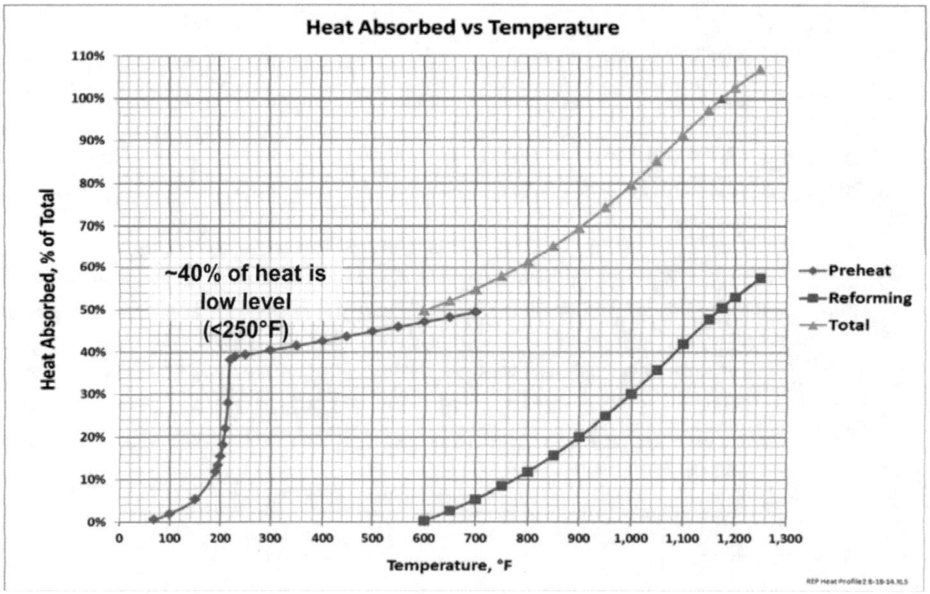

16

FuelCell Energy
Ultra-Clean, Efficient, Reliable Power

Potential Feeds & Heat Sources

- Feeds
  - Natural Gas
  - ADG
  - Anode exhaust
  - Syngas including biomass gasification
- Heat Sources
  - DFC® or other high temperature fuel cell
  - Engines / Turbines
  - Gasifiers
  - Solar collectors
  - Nuclear
  - Steel production
  - Excess electricity
- Potential Advanced Configurations
  - CO_2 Capture may be feasible
    - Burn CO_2/O_2 by-product to produce CO_2/H_2O
  - Scale down to home H_2 production
  - Potential to load follow with low temperature (PEM) fuel cell

17

FuelCell Energy
Ultra-Clean, Efficient, Reliable Power

Capacity of DFC-H2® Plants

DFC300®

DFC1500®

DFC3000®

Co-product	DFC300® CHHP	DFC300® Future REP	DFC1500® CHHP	DFC1500® Future REP	DFC3000® CHHP	DFC3000® Future REP
Power, kW	**250**	210	**1,170**	975	**2,350**	1,950
Hydrogen, kg/day	**135**	300	**635**	1,500	**1,270**	3,000
Heat, mmBtu/hr	**0.5**	0	**2.0**	0	**4.0**	0
Peaker Capacity (Stationary Fuel Cells)						
Peak Power (8 hrs/day), kw	**500**	800	**2,400**	3,900	**4,800**	7,800
Refueling Capacity						
Fuel Cell Cars, 0.5 kg/day	**270**	600	**1,270**	3,000	**2,540**	6,000

Industrial Users of H_2 and Reducing Gas

Heat Treatment, Glass, Refinery, Chemicals,

18

FuelCell Energy
Ultra-Clean, Efficient, Reliable Power

Why SOFC Energy Storage

- Solid Oxide Fuel Cells (SOFC) have demonstrated capability to run in electrolysis mode in addition to fuel cell mode
- Solid Oxide Electrolysis Cells (SOEC) run much more efficiently than conventional electrolysis at high current densities
- Potential for low cost, high efficiency electrolysis combined with high efficiency fuel cell operation implies that SOFC/EC combined with Hydrogen storage may be attractive energy storage approach

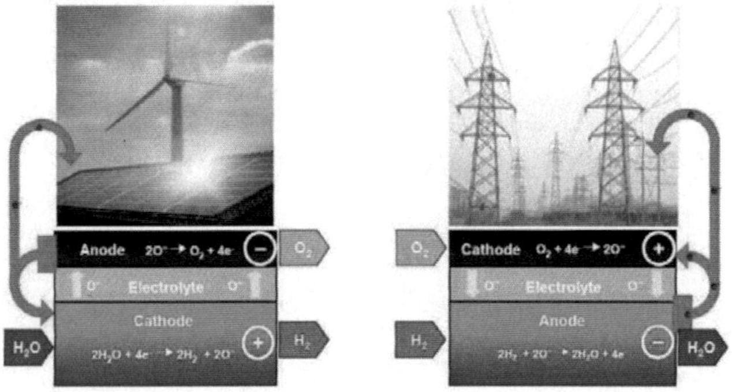

19

FuelCell Energy
Ultra-Clean, Efficient, Reliable Power

FCE/VPS SOFC Technology

- Planar anode-supported cells, 25 cm x 25 cm (550 cm^2 active area)
- Cross-flow gas delivery
- Ferritic stainless steel sheet metal interconnect
- Stacks can be integrated into various arrangements depending on application
- Reversible operation demonstrated: electrolysis and fuel cell mode

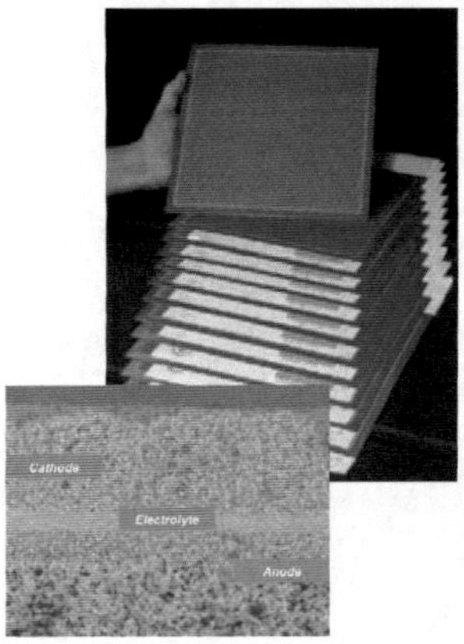

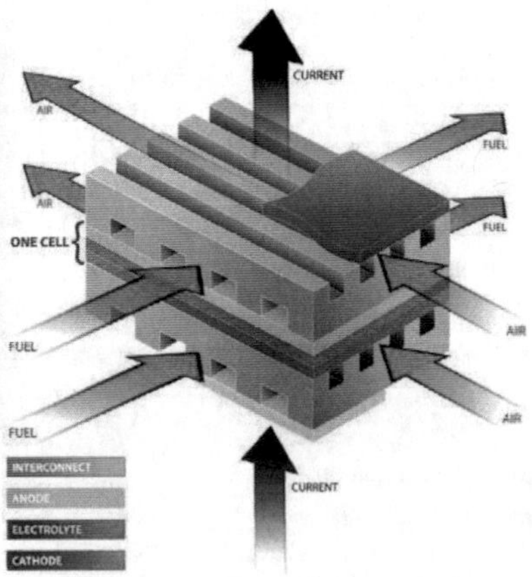

20

SOFC Applications

FuelCell Energy
Ultra-Clean, Efficient, Reliable Power

- Stationary energy storage, targeting subMW markets with 200kW system
- Specialty applications for smaller systems, such as energy storage system in DARPA Vulture unmanned aircraft
- Stationary electrolysis and energy storage

50kW Powerplant

200kW Powerplant

Vulture Unmanned Aircraft, with reversible Solid Oxide stacks in hydrogen based storage system

21

High Efficiency Electrolysis

FuelCell Energy
Ultra-Clean, Efficient, Reliable Power

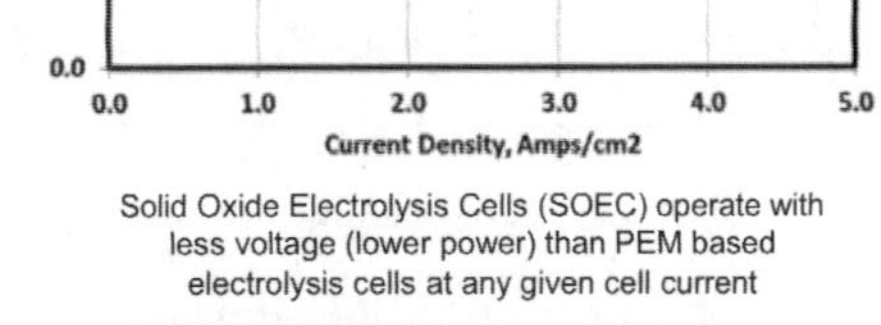

Solid Oxide Electrolysis Cells (SOEC) operate with less voltage (lower power) than PEM based electrolysis cells at any given cell current

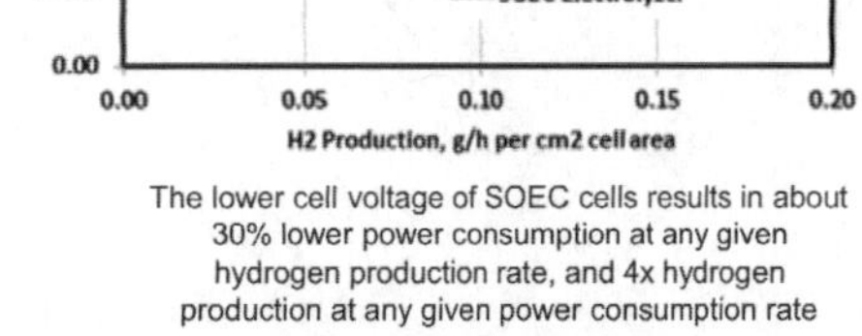

The lower cell voltage of SOEC cells results in about 30% lower power consumption at any given hydrogen production rate, and 4x hydrogen production at any given power consumption rate

In electrolysis mode, Solid Oxide Electrolysis Cells (SOEC) can operate with 30% less power requirement compared to PEM electrolysis (at same production rate), or can produce 4 times as much hydrogen at same cell power consumption rate

22

Reversibility and Efficiency

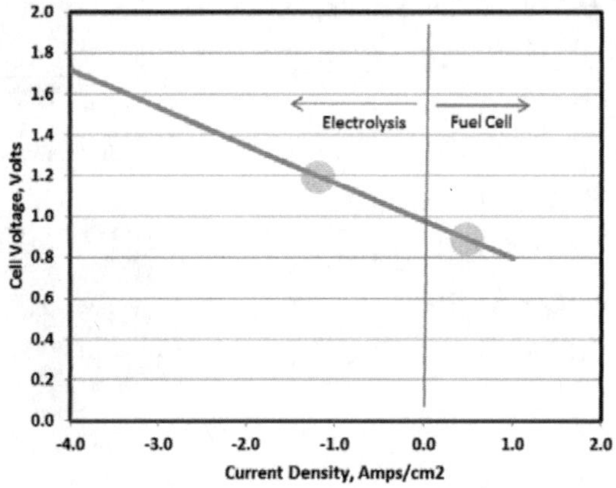

High efficiency electrolysis combined with high efficiency power generation forecasted to achieve high round trip efficiency

23

Discharge Time Drives Cost

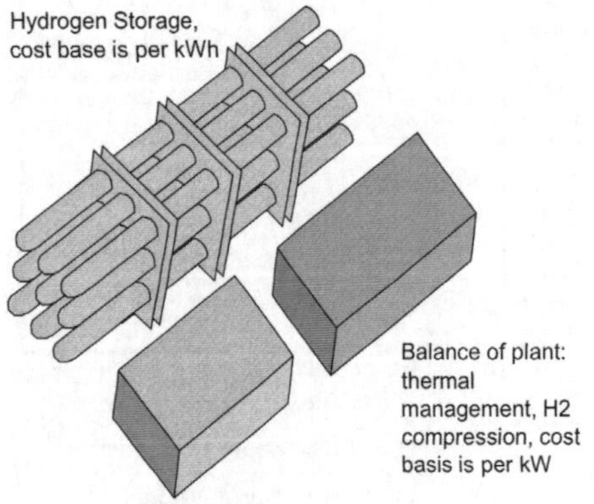

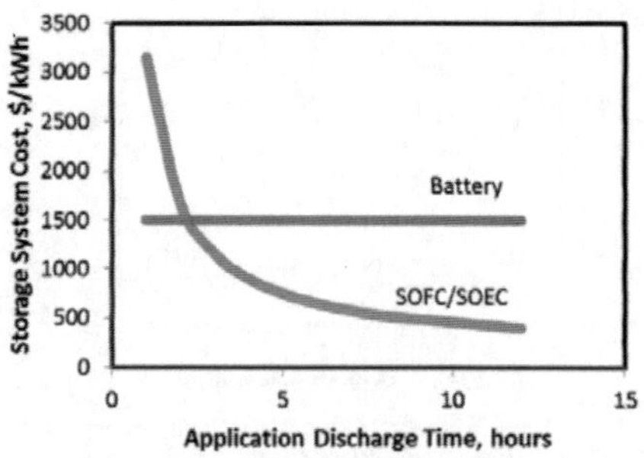

Effect of Discharge Time

SOFC based energy storage has advantages over batteries for long storage duration – storage capacity is increased by adding H_2 storage only

24

FuelCell Energy
Ultra-Clean, Efficient, Reliable Power

Thank you!

Pinakin Patel
Director Advanced Technology
ppatel@fce.com

203-825-6072

25

FuelCell Energy
Ultra-Clean, Efficient, Reliable Power

26

HYSTER-YALE MATERIALS HANDLING

NACCO MATERIALS HANDLING GROUP

About Hyster-Yale | About NACCO Materials Handling | ...tor Relations | Corporate Governance | News Room

News Release

View printer-friendly version

DECEMBER 18, 2014: HYSTER-YALE MATERIALS HANDLING ACQUIRES NUVERA FUEL CELLS

Hyster-Yale Materials Handling, Inc. Announces Acquisi

CLEVELAND, Dec. 18, 2014 /PRNewswire/ -- Hyster-Yale
Inc. ("NMHG"), (collectively, the "Company") acquired Nuvera

Nuvera, located in Billerica, Massachusetts, is a development-stage technology and product
supporting on-site hydrogen production and dispensing systems that can deliver clean energy solutions to custo
technology offers the Company the opportunity for direct access to fuel cells particularly suitable for lift truck applications, as
applicable to its product lines.

Following the acquisition of Nuvera, the Company intends to commercialize Nuvera's research and technology through the rapid integration of this fuel-cell technology across large parts of the Company's lift truck product range. The Company expects to be able to offer its Hyster® and Yale® customers an integrated, factory-fitted fuel-cell solution as well as associated hydrogen generation and delivery capability. In addition, the Company will offer aftermarket solutions designed to fit most any electric powered lift truck brand in the market today. The Company will market these solutions in partnership with its leading independent distribution, national account sales teams and fleet management services to existing and new Hyster® and Yale® customers.

Over the past few years, the Company has been focused on increasing unit volume through market share gains by delivering lowest cost of ownership to its customers in a wide range of applications. In this context, the Company has been evaluating and investing in a broad range of alternative power sources for its lift trucks. The purchase of Nuvera is a strategic acquisition which is expected to provide a platform to expand the Company's offering of best-in-class energy solutions to customers.

While Nuvera technologies have proven capabilities, commercialization of products that utilize these technologies remain in the development stage. As a result, Nuvera is expected to generate significant operating losses over the next two to three years before it can fully commercialize the technology and achieve breakeven results. The acquisition was completed for a modest purchase price with certain contingency payments that are expected to be paid to the selling shareholders based on future deployment of certain elements of the acquired technology. In addition, Hyster-Yale expects to expense up to $40 to 50 million over the next two to three years for additional research and development to commercialize the technology and reach break-even.

The Company believes it is the first major lift truck company to commit to the full deployment of fuel cell solutions for its customers. The acquisition of Nuvera creates a unique capability to integrate fuel cell with lift truck in a way which optimizes the performance and energy efficiency of the combined system, in conjunction with the ability to provide full life cycle maintenance, service and fueling requirements, to meet customers' needs and offer a low overall cost of ownership alternative.

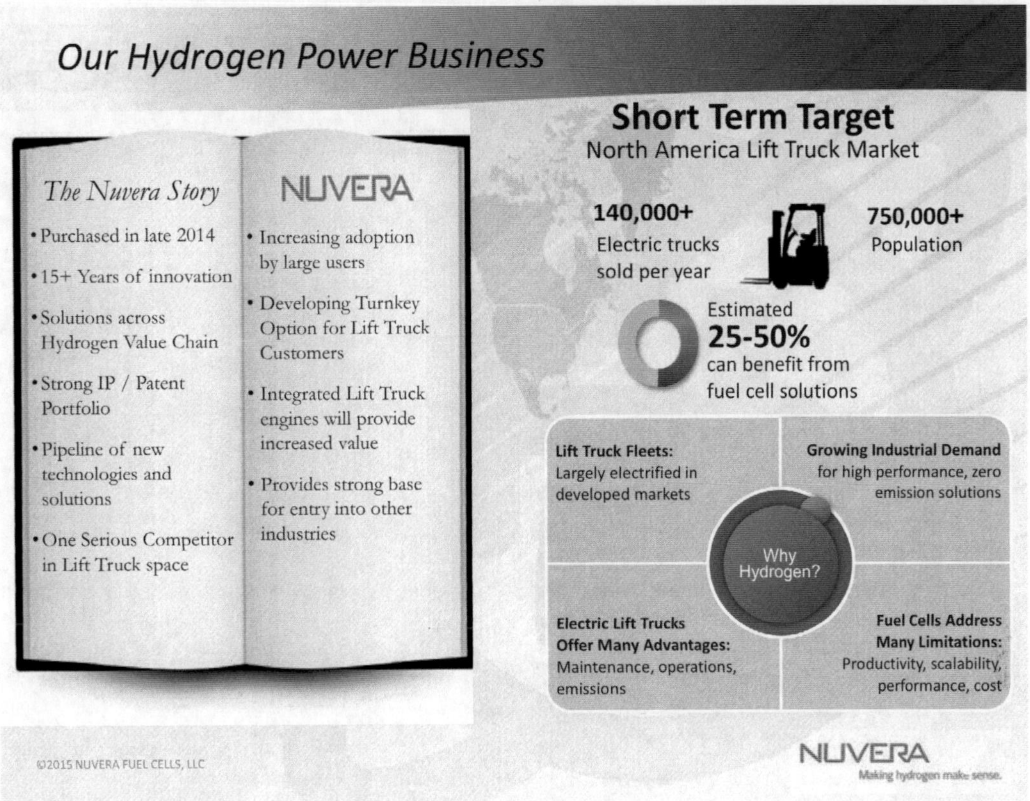

Our Hydrogen Power Business
The Nuvera Story
• Purchased in late 2014
• 15+ Years of innovation
• Solutions across Hydrogen Value Chain
• Strong IP / Patent Portfolio
• Pipeline of new technologies and solutions
• One Serious Competitor in Lift Truck space
NUVERA
• Increasing adoption by large users
• Developing Turnkey Option for Lift Truck Customers
• Integrated Lift Truck engines will provide increased value
• Provides strong base for entry into other industries
Short Term Target
North America Lift Truck Market
140,000+
Electric trucks sold per year
750,000+
Population
Estimated
25-50%
can benefit from fuel cell solutions
Lift Truck Fleets: Largely electrified in developed markets
Growing Industrial Demand for high performance, zero emission solutions
Why Hydrogen?
Electric Lift Trucks Offer Many Advantages: Maintenance, operations, emissions
Fuel Cells Address Many Limitations: Productivity, scalability, performance, cost
©2015 NUVERA FUEL CELLS, LLC
NUVERA
Making hydrogen make sense.

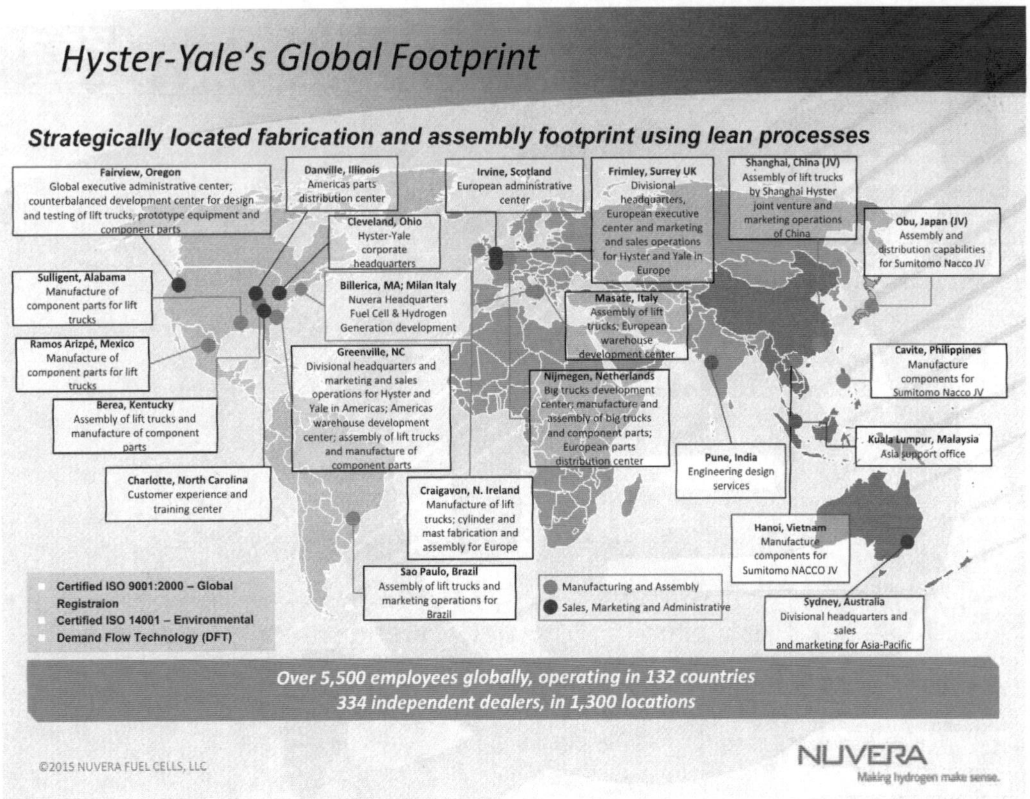

Hyster-Yale's Global Footprint
Strategically located fabrication and assembly footprint using lean processes
Fairview, Oregon
Global executive administrative center; counterbalanced development center for design and testing of lift trucks, prototype equipment and component parts
Danville, Illinois
Americas parts distribution center
Irvine, Scotland
European administrative center
Frimley, Surrey UK
Divisional headquarters, European executive center and marketing and sales operations for Hyster and Yale in Europe
Shanghai, China (JV)
Assembly of lift trucks by Shanghai Hyster joint venture and marketing operations of China
Obu, Japan (JV)
Assembly and distribution capabilities for Sumitomo Nacco JV
Cleveland, Ohio
Hyster-Yale corporate headquarters
Sulligent, Alabama
Manufacture of component parts for lift trucks
Billerica, MA; Milan Italy
Nuvera Headquarters Fuel Cell & Hydrogen Generation development
Masate, Italy
Assembly of lift trucks; European warehouse development center
Ramos Arizpé, Mexico
Manufacture of component parts for lift trucks
Greenville, NC
Divisional headquarters and marketing and sales operations for Hyster and Yale in Americas; Americas warehouse development center; assembly of lift trucks and manufacture of component parts
Nijmegen, Netherlands
Big trucks development center; manufacture and assembly of big trucks and component parts; European parts distribution center
Cavite, Philippines
Manufacture components for Sumitomo Nacco JV
Berea, Kentucky
Assembly of lift trucks and manufacture of component parts
Kuala Lumpur, Malaysia
Asia support office
Pune, India
Engineering design services
Charlotte, North Carolina
Customer experience and training center
Craigavon, N. Ireland
Manufacture of lift trucks; cylinder and mast fabrication and assembly for Europe
Hanoi, Vietnam
Manufacture components for Sumitomo NACCO JV
Sao Paulo, Brazil
Assembly of lift trucks and marketing operations for Brazil
Certified ISO 9001:2000 – Global Registraion
Certified ISO 14001 – Environmental
Demand Flow Technology (DFT)
Manufacturing and Assembly
Sales, Marketing and Administrative
Sydney, Australia
Divisional headquarters and sales and marketing for Asia-Pacific
Over 5,500 employees globally, operating in 132 countries
334 independent dealers, in 1,300 locations
©2015 NUVERA FUEL CELLS, LLC
NUVERA
Making hydrogen make sense.

Core Strategies

COMMITTED to GROWTH

Provide Low Cost of Ownership

Understand Customer Needs

Improve Warehouse Position

Enhance Big Truck Position

Enhance Independent Distribution

Succeed in Asia

Nuvera® Commercialization

NACCO MATERIALS HANDLING GROUP

Why Go Electric?

AC electric drive lift trucks provide performance levels comparable with internal combustion engine trucks...

...and more:

- Lower maintenance costs
- Energy recovery through regenerative braking
- Can be used in- or outdoors
- Use in cold storage
- Zero exhaust emissions
- Greater efficiency

NUVERA
Making hydrogen make sense.

Hydrogen and Fuel Cells for Electric Trucks

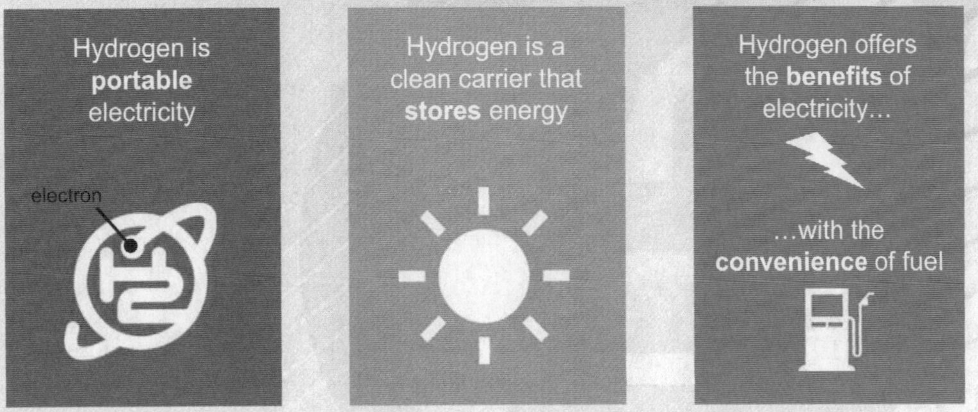

NUVERA
Making hydrogen make sense.

Fuel Cells vs. Batteries in Lift Trucks

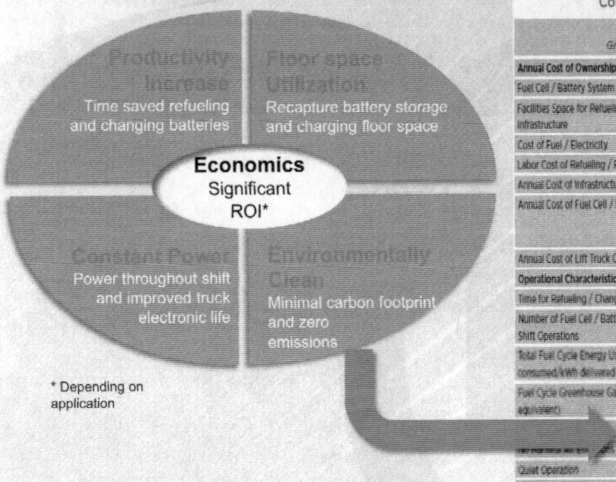

* Depending on application

Government and independent studies confirm lower cost of ownership

Comparison of PEM Fuel Cell- and Battery-Powered MHE[3]

Green Text = Advantage	10kW Class I Forklift		3kW Class III Pallet Jack	
	Fuel-Cell Powered	Battery-Powered	Fuel-Cell Powered	Battery-Powered
Annual Cost of Ownership Per Lift (Total)	$17,800	**$19,700**	$11,700	**$12,400**
Fuel Cell / Battery System Maintenance	$2,200	**$3,600**	**$500**	**$400**
Facilities Space for Refueling / Recharging Infrastructure	$500	**$1,900**	$500	**$1,900**
Cost of Fuel / Electricity	**$2,400**	$500	**$1,400**	$400
Labor Cost of Refueling / Recharging	$800	**$4,400**	$500	**$3,200**
Annual Cost of Infrastructure Capital & Maintenance	**$3,700**	$1,400	**$3,700**	$1,300
Annual Cost of Fuel Cell / Battery Systems	**$2,600** ($3,700 w/o tax credit)	$2,300	**$1,300** ($1,800 w/o tax credit)	**$1,300**
Annual Cost of Lift Truck Capital & Maintenance	**$5,600**	**$5,600**	**$3,900**	**$3,900**
Operational Characteristics				
Time for Refueling / Changing Batteries	6-8 min/day	**30-45 min/day**	3-5 min/day	**25-35 min/day**
Number of Fuel Cell / Battery Systems for Multiple Shift Operations	1	**2-3**	1	**2-3**
Total Fuel Cycle Energy Use (total energy consumed/kWh delivered to the wheels)	~12,000 Btu/kWh	**>14,000 Btu/kWh**	~12,000 Btu/kWh	**>14,000 Btu/kWh**
Fuel Cycle Greenhouse Gas Emissions (g CO_2 equivalent)	800 g/kWh	**1,200 g/kWh**	800 g/kWh	**1,200 g/kWh**
[illegible]	8-10 years	**4-5 years**	8-10 years	**4-5 years**
[illegible] at Point of Use	✓	✓	✓	✓
Quiet Operation	✓	✓	✓	✓
Wide Ambient Operating Temperature Range	✓	✓	✓	✓
Constant Power Available Over Shift	✓		✓	

The comparison of key performance and cost metrics for fuel cell and battery powered lifts are summarized in the table for Class I and Class III equipment. Based on NREL analysis, fuel cell MHE out-performs battery MHE in maintenance, facility space, refueling, recharging time, and product life, resulting in a lower annual cost of ownership for the equipment. Data based on preliminary analysis of early MHE deployments.

Source: http://energy.gov/sites/prod/files/2014/03/f9/early_ markets_mhe_fact_sheet.pdf

NUVERA
Making hydrogen make sense.

Lift Truck Market Opportunity

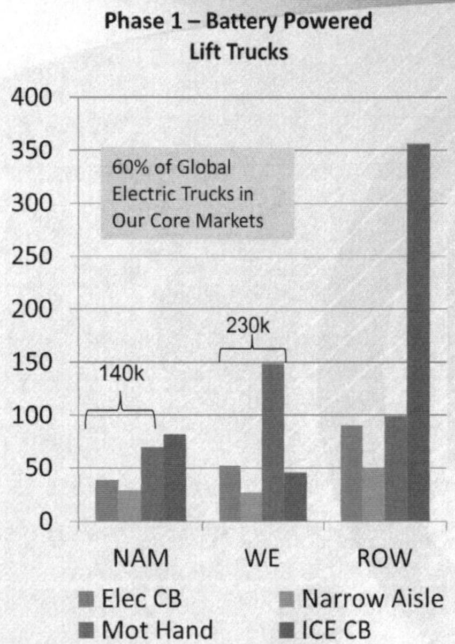

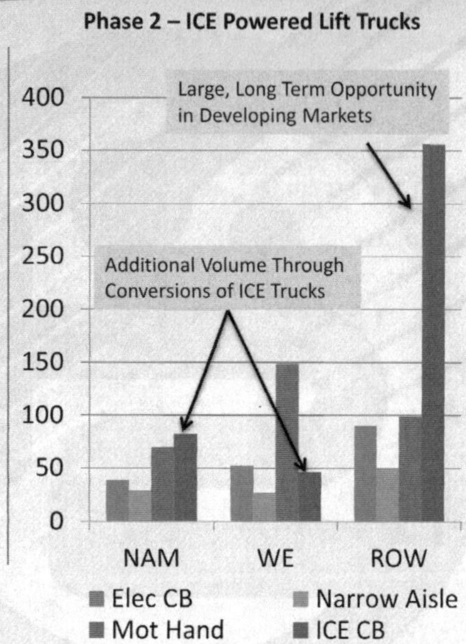

NUVERA
Making hydrogen make sense.

Nuvera Fuel Cells

▲ **Headquarters, Billerica, Massachusetts**
- Executive and administrative offices
- Marketing and sales
- Research and development
- Testing
- Manufacturing

◀ **EU Facilities, San Donato, Italy**
- Fuel cell development
- European sales
- Fuel cell testing

Plus **The Global Strength of HYG**
- Industrial equipment manufacturing, distribution, and service

NUVERA
Making hydrogen make sense.

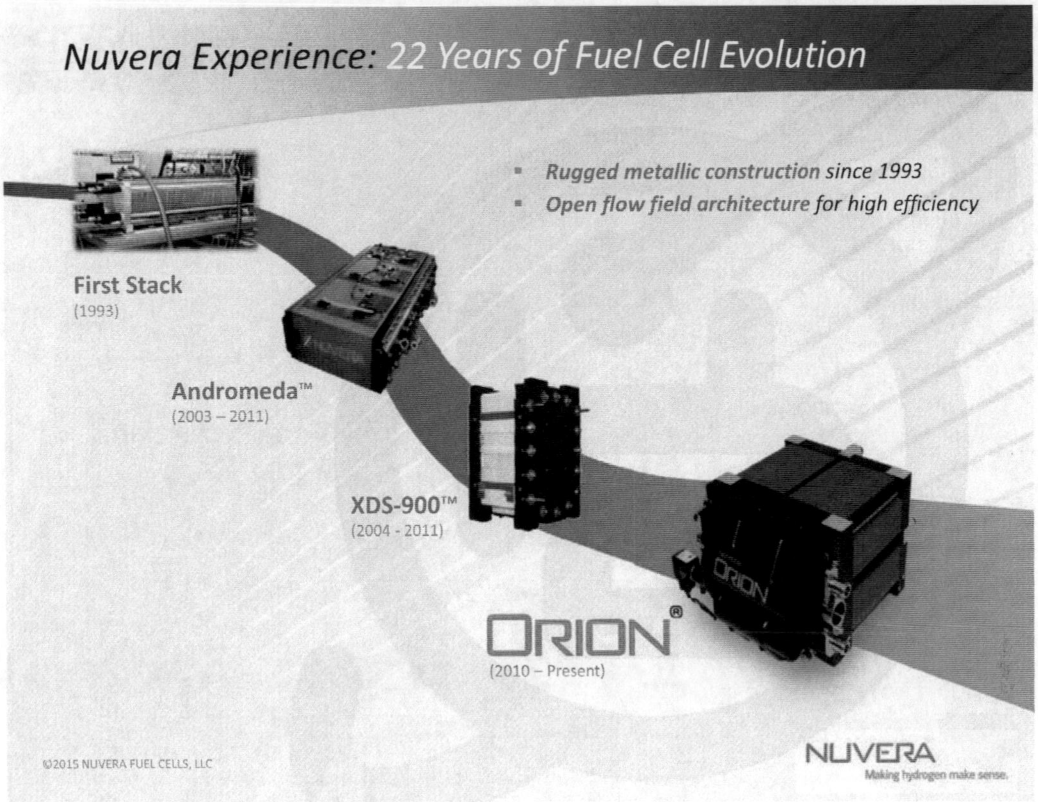

Nuvera Experience: 22 Years of Fuel Cell Evolution
Rugged metallic construction since 1993
Open flow field architecture for high efficiency
First Stack
(1993)
Andromeda™
(2003 – 2011)
XDS-900™
(2004 - 2011)
ORION®
(2010 – Present)
©2015 NUVERA FUEL CELLS, LLC
NUVERA
Making hydrogen make sense.

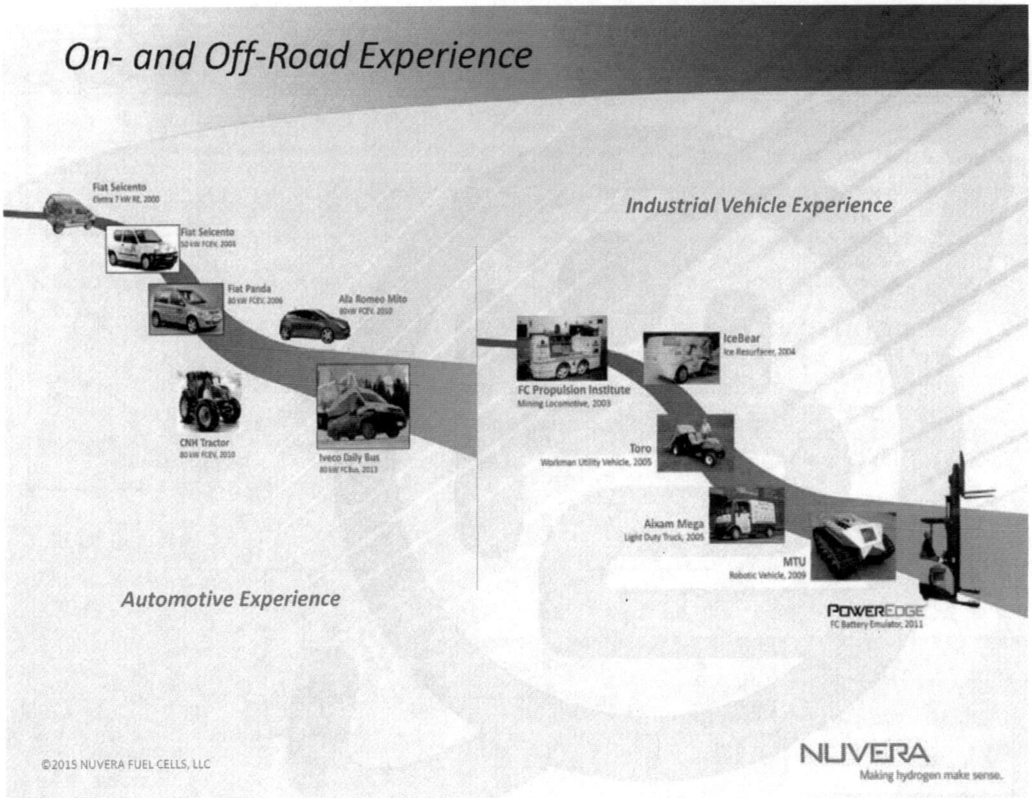

On- and Off-Road Experience
Fiat Seicento
Fiat Seicento
Fiat Panda
Alfa Romeo Mito
CNH Tractor
Iveco Daily Bus
Automotive Experience
Industrial Vehicle Experience
IceBear
Ice Resurfacer, 2004
FC Propulsion Institute
Mining Locomotive, 2003
Toro
Workman Utility Vehicle, 2005
Aixam Mega
Light Duty Truck, 2005
MTU
Robotic Vehicle, 2009
POWEREDGE
FC Battery Emulator, 2011
©2015 NUVERA FUEL CELLS, LLC
NUVERA
Making hydrogen make sense.

Industrial Truck Rollout

Phase I: 2015/16

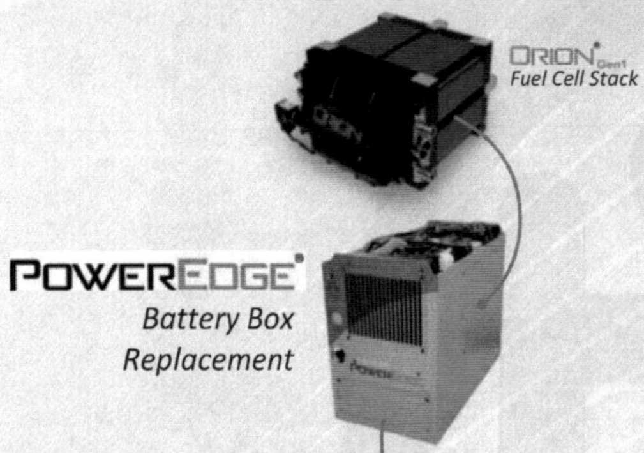

Provides:

- Productivity
- Safety Features
- Emissions Reductions
- Lower Cost of Ownership

NUVERA
Making hydrogen make sense.

Industrial Truck Rollout

Phase II: 2017

- Additional Cost Reduction
- Greater Performance
- Higher Efficiency
- Eliminate Diesel Emissions

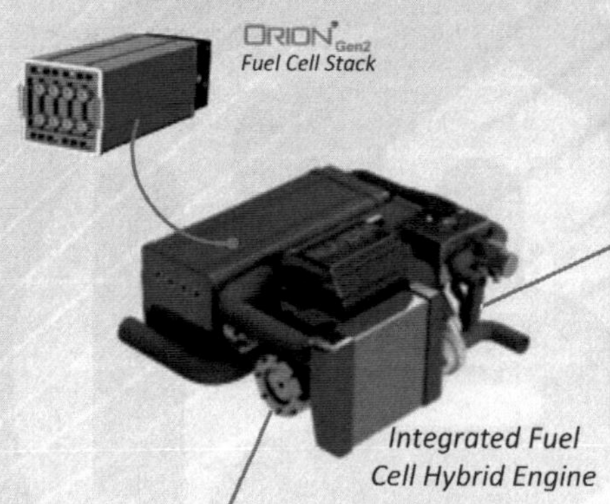

NUVERA
Making hydrogen make sense.

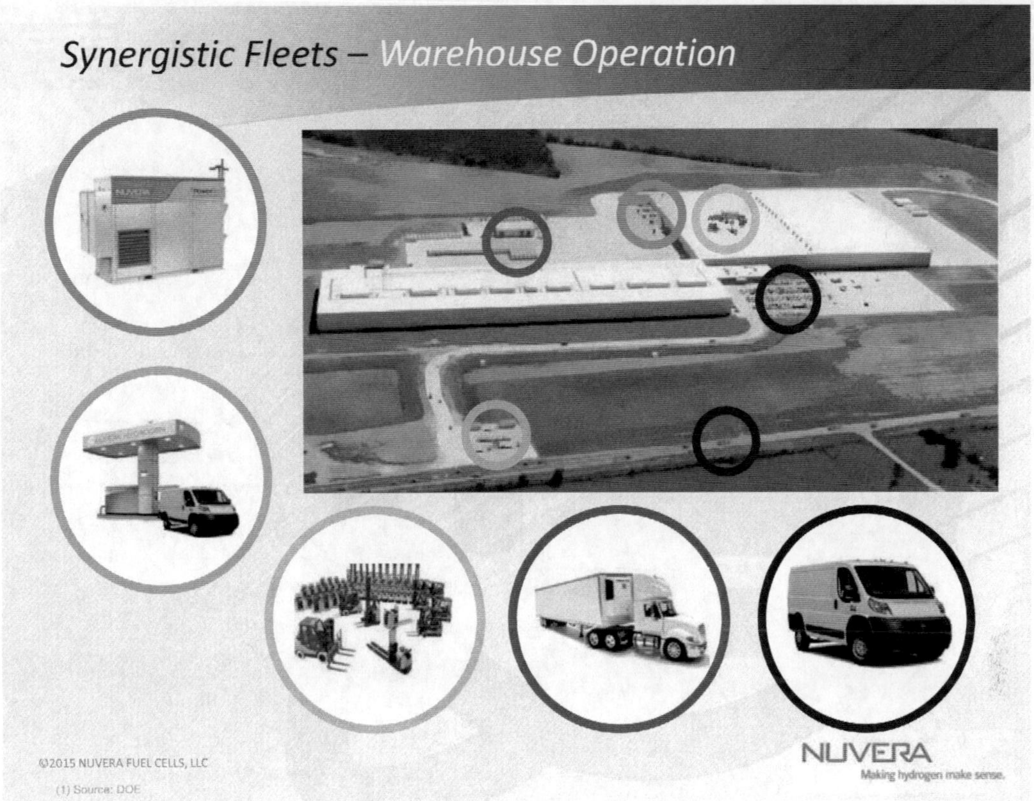
Synergistic Fleets – Warehouse Operation
©2015 NUVERA FUEL CELLS, LLC
(1) Source: DOE
NUVERA
Making hydrogen make sense.

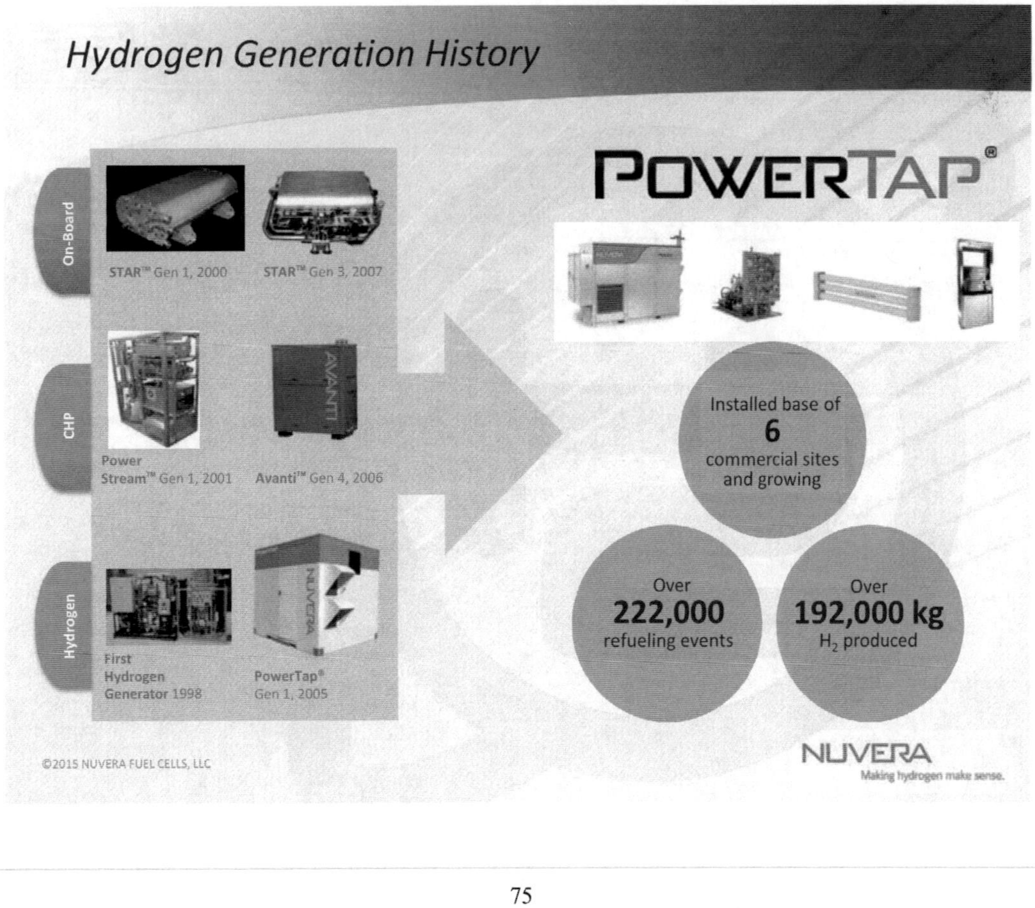
Hydrogen Generation History
On-Board
STAR™ Gen 1, 2000
STAR™ Gen 3, 2007
CHP
Power Stream™ Gen 1, 2001
Avanti™ Gen 4, 2006
Hydrogen
First Hydrogen Generator 1998
PowerTap® Gen 1, 2005
POWERTAP®
Installed base of
6
commercial sites and growing
Over
222,000
refueling events
Over
192,000 kg
H_2 produced
©2015 NUVERA FUEL CELLS, LLC
NUVERA
Making hydrogen make sense.

Delivered Hydrogen vs. On-Site Generation

Delivered Hydrogen

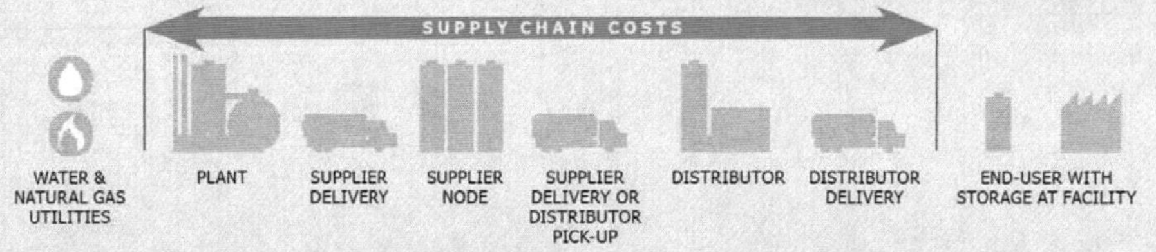

On-Site Hydrogen Generation with PowerTap®

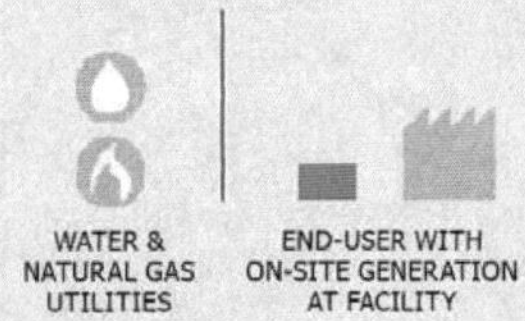

- Eliminate supply delivery costs and related logistics
- Reduce greenhouse gas emissions by eliminating trucking
- Bypass need for storage equipment contracts
- Minimize building offset requirements
- Secure dedicated supply

NUVERA
Making hydrogen make sense.

All-In Hydrogen Fleet Solutions

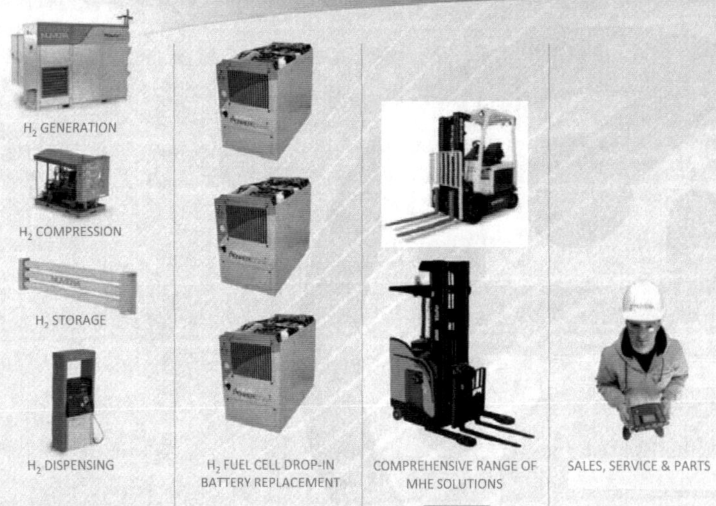

POWERTAP®

POWEREDGE®

HYSTER

Yale

ADDITIONAL HYDROGEN SUPPLY OPTIONS AVAILABLE

HYSTER® YALE®

- Dealer Network
- Financial Services
- Fleet Management
- Innovations
- Rental
- Replacement Parts
- Operator Training

NUVERA®

- Alternative & Redundant Back-Up Hydrogen Supply Options
- Site Planning & Implementation
- Permitting
- H_2 Management

NUVERA
Making hydrogen make sense.

NUVERA
FUEL CELLS
www.nuvera.com/resources
HYSTER
Yale